X Games

in Mathematics

Sports Training That Counts!

Problem Solving in Mathematics and Beyond

Print ISSN: 2591-7234
Online ISSN: 2591-7242

Series Editor: Dr. Alfred S. Posamentier
Distinguished Lecturer
New York City College of Technology - City University of New York

There are countless applications that would be considered problem solving in mathematics and beyond. One could even argue that most of mathematics in one way or another involves solving problems. However, this series is intended to be of interest to the general audience with the sole purpose of demonstrating the power and beauty of mathematics through clever problem-solving experiences.

Each of the books will be aimed at the general audience, which implies that the writing level will be such that it will not engulfed in technical language — rather the language will be simple everyday language so that the focus can remain on the content and not be distracted by unnecessarily sophiscated language. Again, the primary purpose of this series is to approach the topic of mathematics problem-solving in a most appealing and attractive way in order to win more of the general public to appreciate his most important subject rather than to fear it. At the same time we expect that professionals in the scientific community will also find these books attractive, as they will provide many entertaining surprises for the unsuspecting reader.

Published

For the complete list of volumes in this series, please visit www.worldscientific.com/series/psmb

Problem Solving in
Mathematics and Beyond Volume 20

X Games
in Mathematics
Sports Training That Counts!

Tim Chartier

Davidson College, USA

World Scientific

NEW JERSEY · LONDON · SINGAPORE · BEIJING · SHANGHAI · HONG KONG · TAIPEI · CHENNAI · TOKYO

Published by

World Scientific Publishing Co. Pte. Ltd.

5 Toh Tuck Link, Singapore 596224

USA office: 27 Warren Street, Suite 401-402, Hackensack, NJ 07601

UK office: 57 Shelton Street, Covent Garden, London WC2H 9HE

Library of Congress Control Number: 2020051836

British Library Cataloguing-in-Publication Data
A catalogue record for this book is available from the British Library.

Problem Solving in Mathematics and Beyond — Vol. 20
X-GAMES IN MATHEMATICS: SPORTS TRAINING THAT COUNTS!

ISBN 978-981-122-383-9 (hardcover)
ISBN 978-981-122-487-4 (paperback)
ISBN 978-981-122-384-6 (ebook for institutions)
ISBN 978-981-122-385-3 (ebook for individuals)

For any available supplementary material, please visit
https://www.worldscientific.com/worldscibooks/10.1142/11924#t=suppl

Desk Editors: Vishnu Mohan/Tan Rok Ting

Typeset by Stallion Press
Email: enquiries@stallionpress.com

To my big sis Melody.
You invited me into the game
as your teammate, opponent, and coach.
Thank you for being my fan
in the game of life!

Preface

I have played sports for much of my life. Athletics filled my childhood days, especially during the summer. My sister, Melody, served as my opponent, teammate, and coach in many of the competitions, although we didn't always keep scores. Growing up, many of my afternoons and weekends consisted of walking over to the nearby parking lot where a group would gather and we'd decide what to play. We would begin playing and iterate the rules until we were satisfied. We played baseball with tennis rackets, basketball with a football, and football with any ball we could toss. One day, a lot of younger children arrived to play; so we assigned teams to keep things as fair as possible, which ended up being a 2 on 10 basketball game. Everyone was welcome. Everyone could play.

I'm now a professor of mathematics and computer science at Davidson College in North Carolina. My scholarship is playing sports — not on the field, court, or track, but with mathematics. Seven years ago, three students walked into my office with a simple request: "Can we form a group that would supply analytics to the Davidson College men's basketball team?" Today, what began with three students now involves almost 100. The group, now known as "Cats Stats", supplies analytics to our Division I college coaches in men's basketball, women's basketball, women's soccer, men's soccer, volleyball, field hockey, swimming, and football. Our success has led to projects with professional teams in the NBA, NFL, and NASCAR. We've fielded analytics questions from *The New York Times*, *ESPN*, and the *Washington Post*. We have a simple principle: If someone wants to be involved, we find a place. Everyone can play.

You've picked up this book, so I presume you have an interest and want to "play" sports analytics. Writing this book, I've imagined myself collaborating with students from my sports group, engaging students in the class, or speaking to 50 or 1,000 people. The goal is simple: to get you into the game of math and sports. You have a place.

Do you like sports? This book will detail ways of analyzing athletics to gain insight that can otherwise be obscured. The book also underscores how analytics can springboard into other areas of mathematics.

Do you like math? You'll find in this book many mathematical topics not involving sports. You'll also see how sports analytics can train you broadly in mathematics. It is my hope that in these pages you find a place to explore.

For example, in 2009, Dr. Amy Langville of the College of Charleston and I worked with two students to develop ranking techniques for the NCAA Division I basketball tournament, commonly known as "March Madness". The achievement has been covered widely in the media and results in my phone ringing every March to discuss my work. In Chapter 6, you'll learn how to create your own predictions. When I taught the methods in the class, students who loved sports enjoyed using mathematical ideas to study athletics. Students who didn't follow sports commented that the work enabled them to engage in sports conversations in new ways. One student entered her family March Madness pool for the first time — and won rather handily. I've worked to write a book that can engage readers in similar ways for a much broader range of topics.

So, welcome to my office! Come in and let's chat. I'll share my knowledge. It is a hope that this book will spark your ideas. I'd love to know what you create as I'll learn from you. Let's begin!

About the Author

Tim Chartier is a professor of mathematics and computer science at Davidson College and specializes in sports analytics. He consults with ESPN, the *New York Times*, the US Olympic Committee and teams in the NBA, NFL and NASCAR. He oversees close to 100 student researchers who supply analytics to Davidson College sports teams, which includes basketball, football, soccer, volleyball, baseball and swimming. Tim was the first chair of the Advisory Council for MoMath. He has also worked with Google and Pixar on their educational initiatives. He has received a national teaching award from the Mathematical Association of America (MAA). His research and scholarship were recognized with an Alfred P. Sloan Research Fellowship. Through the Teaching Company, Tim completed a 24-lecture series entitled Big Data: How Data Analytics Is Transforming the World. Tim authored the book When Life is Linear: From Computer Graphics to Bracketology which won the MAA's Beckenbach Book Prize. He wrote Math Bytes: Google Bombs, Chocolate-Covered Pi, and Other Cool Bits in Computing which won the Euler Book Prize. Tim also coauthored the textbook Numerical Methods: Design, Analysis, and Computer Implementation of Algorithms. The MAA recognized the impact of Tim's writing on undergraduate mathematics education with the Daniel Solow Author's award.

Acknowledgments

Before discussing analytics, I want to express my gratitude to a variety of people. First and foremost, I thank my wife Tanya Chartier. Tanya, thank you for encouraging me while I worked on this book. At different stages, I could sense the book I would write but a clear view was beyond my gaze. I see you in so many pages of this book, which brings even greater joy to my authorship. I thank my children, Noah and Mikayla, for listening to me to share exciting examples and for supporting my work on sports.

This book is dedicated to my sister, Melody. Mel, I appreciate your consent to read and give feedback on my mathematical writing. Your encouragement has often helped me to tune my writing to the audiences who don't enjoy math.

MaryJo Johnson, thank you for listening to me discuss my book, and my hopes for what it could be. I trust that you will find that this book is a realization of those dreams.

Mom and Dad, thank you for supporting my wide interests. I'm a mathematical mime who works on sports. You've always enjoyed watching me continue to discover who I'd be. There has always been room in my life to further "become". I enjoy sharing life with you as it unfolds, which isn't always easy but can bring great meaning.

I'm grateful to my mother-in-law Carolyn Harman. The very first version of this book came from a summer we spent large amounts of time together. I sense her smile and laughter in these pages. I miss you, Mom.

I'm grateful to my editor Rochelle Kronzek. Shelley, thank you for working with me while this book took many forms. I'm proud of what we've produced and can vividly remember the moment you shared your vision for this version of the book. I sat entirely still as ideas flooded my mind. This book stems from that silence. Thank you for your patience and encouragement.

Looking at the bibliography, it is clear that I've coauthored a number of articles. Many of those are with undergraduates. As I sit here now, a quick computation indicates that I've worked with several hundreds of students in sports projects. I don't know how to thank them all. I want to thank George Baldini for his work on NBA analysis as shown in Section 3.2. Thanks go to Tife Odumosu for his cartoon in Figure 2.9. Thanks also go to Ansley Earle for her artwork in Figure 18.7 and her illustrations for the solutions to the puzzle at the end of the chapter. I'm grateful to the innovative group of students who took my Linear Algebra in Data Mining & Computer Graphics course at Davidson College for creating some of the correlations in Chapter 5.

Thanks go to Davidson College. I appreciate those students who repeatedly affirmed my work; you helped me to dream bigger and reach farther. I appreciate the faculty and staff, who surround me with excellence and dedication. I also thank Jay Pfeifer, who absorbed and encouraged my enthusiasm for math and sports; thanks for the many ways you've aided my work.

I'm grateful to the participants of my sports and math seminar at the Charlotte Teachers Institute. Public school teachers in the seminar inspired me by their curricular goals. Some of the content of this book came from that experience, and others springboarded from them. Thanks go to Michael Pillsbury for serving as the seminar coordinator; you are simply a great teacher, collaborator, and friend.

Thanks go to colleagues who collaborate with me in sports research. Amy Langville is a friend and collaborator extraordinaire. My colleagues at Furman University demonstrate how teaching and research enrich each other; thanks go to Liz Bouzarth, John Harris and Kevin Hutson. I'm also grateful to Ben Baumer, Rick Cleary, Russ Goodman, Roland Minton, Megan Powell, and Debby Stroman — each of you inspire me in your work and its impact on student lives. Thanks go to Ken Massey for fielding my many questions through the years. I appreciate Peter Keating of ESPN; sitting next to you is always a brainstorming hurricane that I enjoy. I'm grateful to Tresata and Abhi Mehta for the opportunity to explore their predictive analytics tools in the context of sports. Thanks go to Tom Moore of

Grinnell College for his note that directed me to the transportation example of Simpson's paradox. Finally, thanks go to the National Amateur Sports and James Shipley for the opportunity to develop a sports analytics program for secondary students.

It is impossible to thank everyone who were involved in the project and supported its content. For those unmentioned here, may you see your influence in these pages and enjoy the way our interaction materialized into printed work. So, let's get into the game!

Contents

Chapter 1

Game On!

Numbers play a foundational role in sports. Babe Ruth hit 714 career home runs, a record that would hold for almost 40 years. Florence Griffith-Joyner set a world record in the 100 meters by sprinting from start to finish in 10.49 seconds in 1988, which remained a record for decades. Bob Beamon set the long jump world record by bounding 29 feet 2 and 1/4 inches in 1968 beating the farthest previous leap by almost 2 feet. The University of Connecticut women's basketball team won 111 consecutive games between 2014 and 2017, which included two national championships.

Sports are more than scores. The numbers can tell us more than records. The numbers and statistics of sports can also inform strategic decisions. For example, consider baseball, a sport with a longer tradition of using numerics than many sports. Innovations still occur but are not always as new as one might think. Let's go back to the 1940s for such an example. From 1939 to 1946, Ted Williams, one of the best hitters of baseball, saw defenses much like Figure 1.1(a). Then, on July 14, 1946, Ted Williams walked to the plate in the second game of a doubleheader to see the Cleveland Indians' defense arranged much like the configuration in Figure 1.1(b) [48, 52]. Williams was reported to laugh at the sight. Then, he hit into the shifted defense and was thrown out by the shortstop, who was standing between the first and the second base. The shortstop was Lou Boudreau, who was also the manager as this occurred at a time when players could play and manage their team. Boudreau broke out a defensive strategy that was only sparingly used before

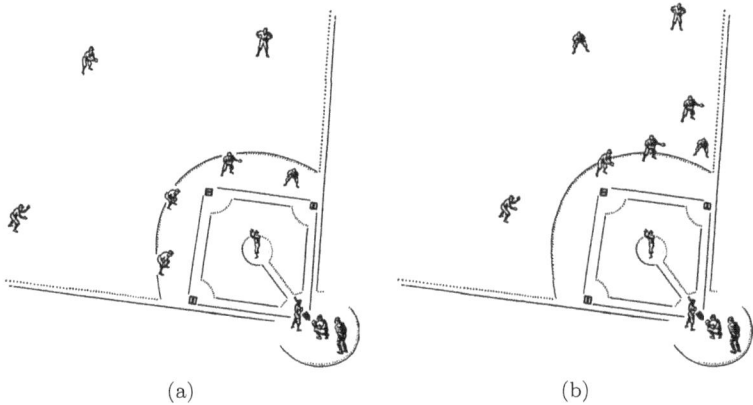

(a)   (b)

Figure 1.1: Before mid-July 1946, Ted Williams faced defenses as seen in (a). After one game, defenses began shifting to where Williams typically hit the ball. His first defensive shift had a configuration much like (b).

in baseball history — the shifted defense. Williams would often face some form of a shifted defense from that day forward. And today, many teams use a shifted defense with much attention being paid to the strategy as it, in part, employs percentages. If you can cover where a player will hit the ball, then you are better protected from a hit or, at the least, a hit leading to extra bases.

Using math to study sports is more than just dabbling in mathematical ideas. Sports enable us to venture into many areas of mathematics. So, this book will enable us to step into the X Games, where we study games of sport with math (like algebra where x is a well-known friend) and where sports (and its games) springboard us into various areas of mathematics. So, each chapter has topics from sports, which will be connected in the chapter to mathematical areas far from athletic competitions.

1.1. Numeric Eyes

Beyond counting seconds or a score, why do we use math in sports? To begin answering this, grab a 6-sided die. We'll play baseball, at least simulated baseball, which is a means to understand and dig into the numbers of a game. We'll discuss simulations further in a later chapter.

To begin with, you'll be a player called Sigma. With the roll of a die, you'll determine if there is a hit. If you roll a 1 or 2, Sigma gets a hit. Roll your die 10 times. How many hits did Sigma get?

Now, you'll step to the plate as Delta. Roll the die twice 10 times. This time, if you roll a 1, 2, or 3 on the first roll and a 1, 2, or 3 on the second roll as well, you get a hit. How many hits did you get this time?

Who had more hits? I actually don't know whether you will say Sigma or Delta. I do know that if you had rolled 100 times or, even better, 10,000 times, then my chances of predicting who had more hits improve. But, in only 10 at bats, what I know is that I don't know.

Probability helps us to see the underlying long-term behavior. The probability that Sigma will hit the ball is 2/6 since two outcomes of rolling the die produce a hit and there are six total possible outcomes. Therefore, Sigma is constructed to have a 2/6 probability of a hit, which directly gives a batting average of 0.333. In baseball, a batting average is calculated as the number of hits divided by the official number of at bats and is expressed as a decimal to three places of accuracy. Note that if you achieved a 0.333 batting average over a career, you'd be in the top 25 Major League Baseball hitters of all time!

How about Delta's batting average? There are nine outcomes that have 1, 2 or 3 on the first roll and a 1, 2 or 3 on the second roll, so Delta's batting average is $9/36 = 0.250$. This falls below the league average from 1973 to 2017.

This is where we see the roll of numbers, statistics, and analytics. A Major League Baseball (MLB) player has around 550 at bats during a season. We'd expect Sigma to have approximately 550 × 0.333 or about 183 hits. Delta would have approximately 138 hits. So, Sigma would have about 45 more hits than Delta. There are 162 games in a MLB season. The difference between Sigma and Delta equates to 1 more hit every 3.6 games. This is about 1–2 more hits per week, since approximately 6 games are played per week in MLB.

Having 1 or 2 more hits per week is the difference between being below average and one of the best hitters of all time. Batting 0.333 versus 0.300 corresponds to less than 1 hit per week. The numbers, 0.250, 0.300 and 0.333, help discern these differences at play. Our understanding of the numbers help us to glean such evaluations of quality.

1.2. Story Behind the Numbers

Numbers create a skeleton outlining the events of a sporting contest. Let's take an extreme example in basketball. Did you hear about the guard in NCAA basketball that LeBron James compared to the likes of Kobe Bryant and Wilt Chamberlain? That comparison came in 2012 during the player's sophomore year. On November 20, 2012, Jack Taylor from Grinnell College scored an unbelievable 138 points, breaking a 58-year-old NCAA record for points in a single game. Taylor would score 109 points in November 2013. Grinnell won both games. They scored 179 points in the 2012 game and 173 in the 2013 contest.

How can one player shoot over 100 points in a college basketball game, let alone do so twice in less than a year? We will get a bit of the picture if we look at the other records set in the game. Taylor broke NCAA records for field goals (52), field goals attempted (108), three-pointers (27), and three-pointers attempted (71). He also set a record for Division 3 schools by scoring 28 consecutive points.

That's a lot of scoring. Some quick computations uncover the nature of that game, even if you didn't travel to Iowa to watch it. In his 138-point performance, Taylor played 36 of the 40 minutes in the game and made 52 of his 108 shots (see Figure 1.2). This equates to three shots a minute, and 70 of those shots were three-point attempts. This means that Taylor averaged two three-point attempts per minute.

Figure 1.2: Jack Taylor of Grinnell College who scored 138 points in one game.

A modern basketball statistic computes a team's efficiency while a player is on the floor. To examine the efficiency, the number of possessions that a team has in a game is required. However, possessions are not recorded in traditional box scores. Now, a team possession happens every time a player of that team (1) attempts a field goal, (2) misses a shot that does not lead to an offensive rebound, (3) turns the ball over, or (4) shoots free throws and either makes the last shot or does not result in an offensive rebound from a missed last shot. We can work with box scores to get an estimate of these events. So, one estimate of the number of possessions in a game is

$$\text{number of possessions} = (\text{FGA} - \text{OR}) + \text{TO} + 0.44(\text{FTA}),$$

where FGA is the field goal attempts, OR is the offensive rebounds, TO is the turnovers, and FTA is the free throw attempts. The 0.44 multiplier takes into account that not all free throws, like a technical foul shot, involve a possession. John Hollinger, in *Pro Basketball Prospectus*, proposed a multiplier of 0.44, which Ken Pomeroy, on his popular college basketball website kenpom.com, says is the same for college basketball [51].

With this formula, we can use the box scores from the Grinnell games to compute an estimate on the number of possessions in a game. First, Jack's 138-point performance involved 136 field goal attempts, 37 offensive rebounds, 17 turnovers, and 16 free throw attempts. So, the number of possessions is computed as $(136 - 37) + 17 + 0.44(16) \approx 123$. The 2013 game involved 125 field goal attempts, 32 offensive rebounds, 7 turnovers, and 29 free throw attempts. So, the number of possessions in this game is calculated as $(125 - 32) + 7 + .044(29) \approx 101$. From this we see that the game where Jack scored 138 points was played at a faster pace than the 2013 game.

With possessions in hand, we can go another step and compute efficiency, which is calculated as the number of points divided by the number of possessions. So, Grinnell averaged $179/123$ or 1.46 points per possession, denoted PPP, indicating that one would expect Grinnell to score 146 points in 100 possessions. This is a very high efficiency. For the 2013 game, Grinnell averaged $173/101$ or 1.71 PPP, which is an even higher efficiency. Interestingly, Grinnell had a higher efficiency in the 2013 game with a slightly lower score. Points per possession are a way of comparing the performance in games with different paces.

Let's go even further and estimate the team's efficiency when Jack Taylor was on the floor. If he played the entire game in 2012, then the team's 1.46 PPP would represent the team's efficiency while Taylor was on the court. However, he played 36 of 40 minutes or 90% of the game. How many possessions did the team have while Jack was on the court? That's also not easily accessible. Our estimate will be 90% of the team's 123 possessions or 110.7 possessions. In the 2013 performance, Jack played 29 minutes or 72.5% of the game. So, we estimate the number of possessions for Grinnell when Jack was on the floor to be 73.225.

We can also compute the individual offensive rating as 100 (points scored/possessions). In the game Jack scored 138, the offensive rating of Grinnell when Jack was on the court was $100(138/110.7) \approx 125$. In Jack's 109-point game, the offensive rating is $100(109/73.225) \approx 149$. In both games, Taylor was scoring at a ridiculous rate, which caught the attention of NBA players like LeBron James. Yet, somewhat surprisingly, Jack's offensive rating was actually better in his 109-point outing than in his 138-point outing.

Note how these computations folded pace into their calculations. A player with a higher offensive rating might not actually score more points than Jack. We saw how Jack's record of 138 points didn't result from his highest offensive rating. Also, offensive rating is reported relative to a 100-point game. Many NCAA games don't reach 100 points for a team. Further, offensive rating assumes a player can play at a given rate throughout the game, without wearing out. Coaches comment on how players vary in how long they are productive. In other words, players vary in the size of their gas tanks. What about Jack Taylor? The numbers clearly tell us that he had a pretty deep gas tank.

It's notable that Jack Taylor scored over 100 points in two games. When we dig into the numbers, we get an even fuller picture of this record-setting achievement and see more of the story behind the numbers.

1.3. LeBron's Sharp Mathematical Eye

We just saw how box scores can give a synopsis on a game. Box scores capture the individual and team performances in the game. The official scorer records the box score statistics for a game. Baseball has a long history of box scores, with many fans keeping track of the game on their own as they tally events of the game. Keeping the score of a game involves tracking every pitch and hit. Be careful waving down a hot dog vendor as you could miss a play. The good news is there are ways to catch some of the errors.

If you keep score, then you want your box score to be balanced at the end of the game. Pick a team. First, add that team's number of at bats, bases on balls, hit batters, sacrifice bunts, sacrifice flies, and batters awarded first base because of interference or obstruction. Then, find the total of that team's runs, players left on base, and the number of outs made by the opposing team. These two totals should be equal. If they aren't, the box score is unbalanced and has an error.

Box score errors happen, even at the highest level of sports. After losing to the San Antonio Spurs in Game 1 of the 2013 NBA Finals, LeBron James scanned the stat sheet, as seen in Figure 1.3 during a postgame news conference. James spotted an inconsistency and noted, "I was looking at the stat sheet, and it says that they had 21 second-chance points. I don't really understand how that is possible with only six offensive rebounds. I'm very good in math" [9].

Second-chance points describe any points scored during a possession by the offense after an offensive player has already attempted one shot and missed. Note, a second-chance point requires an offensive rebound in order for the offensive team to maintain possession of the ball. LeBron continued doing the math in his head and noted to the media, "The only way you can get a second-chance point is if you can get an offensive rebounds, right? Am I correct? So, even if you hit a 3 off of 6 offensive rebounds, that's still only 18 points."

Errors can be corrected, of course. For one New York Giants football player, such a correction was a bonus — a million dollar bonus. After the December 22, 2019 game against the Redskins, Giants linebacker Markus Golden had 9.5 sacks. During the game, he was credited with half a sack. However, after the game, the stat was upgraded to a full sack. With that, Golden had 10 sacks on the season, which triggered a million dollar bonus from his contract. Fractions matter. For Golden, it was a million dollar difference!

NATIONAL BASKETBALL ASSOCIATION   OFFICIAL SCORER'S REPORT
                                                          FINAL BOX

6/6/2013 American Airlines Arena, Miami, FL
Officials: #13 Monty McCutchen, #23 Jason Phillips, #25 Tony Brothers

Time of Game: 2:29
Attendance: 19,775 (Sellout)

VISITOR: San Antonio Spurs (1-0)

NO	PLAYER		MIN	FG	FGA	3P	3PA	FT	FTA	OR	DR	TOT	A	PF	ST	TO	BS	PTS
2	Kawhi Leonard	F	35:23	3	9	0	4	4	4	2	8	10	1	2	1	0	0	10
21	Tim Duncan	F	36:35	8	19	0	0	4	4	3	11	14	4	4	1	2	3	20
22	Tiago Splitter	C	24:48	3	6	0	0	1	2	1	1	2	0	0	0	1	1	7
4	Danny Green	G	34:08	4	9	4	9	0	0	0	5	5	0	3	1	0	1	12
9	Tony Parker	G	39:32	9	18	0	0	3	4	0	0	0	6	1	1	0	0	21
20	Manu Ginobili		29:31	4	11	2	5	3	4	0	0	0	3	2	2	1	0	13
33	Boris Diaw		8:58	1	1	0	0	0	0	0	1	1	1	0	0	0	0	2
5	Cory Joseph		2:14	0	1	0	0	0	0	0	0	0	0	0	0	0	0	0
14	Gary Neal		21:34	3	9	1	5	0	0	0	2	2	1	0	0	0	0	7
15	Matt Bonner		7:17	0	1	0	0	0	0	0	3	3	0	0	0	0	0	0
45	DeJuan Blair		DNP - Coach's Decision															
1	Tracy McGrady		DNP - Coach's Decision															
8	Patty Mills		DNP - Coach's Decision															
	TOTALS:			35	84	7	23	15	18	6	31	37	16	12	6	4	5	92
	PERCENTAGES:			41.7%		30.4%		83.3%		TM REB: 10					TOT TO: 4 (8 PTS)			

HOME: MIAMI HEAT (0-1)

NO	PLAYER		MIN	FG	FGA	3P	3PA	FT	FTA	OR	DR	TOT	A	PF	ST	TO	BS	PTS
6	LeBron James	F	41:35	7	16	1	5	3	4	2	16	18	10	0	0	2	0	18
40	Udonis Haslem	F	16:43	1	1	0	0	0	0	0	4	4	0	2	0	0	0	2
1	Chris Bosh	C	35:19	6	16	0	4	1	2	1	4	5	1	4	3	1	1	13
3	Dwyane Wade	G	35:38	7	15	0	0	3	4	2	0	2	2	0	1	1	0	17
15	Mario Chalmers	G	27:18	3	10	2	6	0	0	0	1	1	2	0	0	1	0	8
34	Ray Allen		23:57	3	4	3	4	4	5	1	2	3	0	1	0	1	0	13
13	Mike Miller		19:57	2	3	1	2	0	0	0	4	4	0	0	0	1	0	5
11	Chris Andersen		12:58	3	5	0	0	1	2	1	2	3	0	3	0	1	0	7
30	Norris Cole		17:12	2	4	1	1	0	0	0	2	2	4	1	0	0	0	5
31	Shane Battier		6:11	0	3	0	3	0	0	0	1	1	1	1	0	0	1	0
50	Joel Anthony		3:12	0	1	0	0	0	0	2	1	3	0	0	0	0	0	0
22	James Jones		DNP - Coach's Decision															
9	Rashard Lewis		DNP - Coach's Decision															
	TOTALS:			34	78	8	25	12	17	9	37	46	20	12	4	8	2	88
	PERCENTAGES:			43.6%		32.0%		70.6%		TM REB: 8					TOT TO: 9 (15 PTS)			

SCORE BY PERIODS	1	2	3	4	FINAL
Spurs	23	26	20	23	92
HEAT	24	28	20	16	88

Inactive: Spurs - Baynes, De Colo Heat - Howard, Varnado

Pts. in the Pt. Spurs 40 (20/38), HEAT 34 (17/34)   Biggest Lead Spurs 7, HEAT 9
2nd Chance Pts. Spurs 21 (8/10), HEAT 8 (4/7)   Lead Changes 7
FB Pts. Spurs 4 (1/6), HEAT 9 (4/5)   Times Tied 6

Figure 1.3: The official scorer's report from the first game of the 2013 NBA Finals between the Miami Heat and San Antonio Spurs. LeBron James spotted an inconsistency.

1.4. Zero Impossibility

LeBron James had a sharp mathematical eye to spot the error in the stats sheet. Can you spot an error that leads to the impossibility of 0 equalling 1?

$$a = b \qquad \text{our initial identity}$$
$$a(a - b - 1) = b(a - b - 1) \qquad \text{multiplying both sides by } (a - b - 1)$$
$$a^2 - ab - a = ab - b^2 - b \qquad \text{distribution}$$
$$a^2 - ab = ab + a - b^2 - b \qquad \text{adding } a \text{ to both sides}$$
$$a(a - b) = a(b + 1) - b(b + 1) \qquad \text{factoring both sides}$$
$$a(a - b) = (b + 1)(a - b) \qquad \text{another step of factoring}$$
$$a = b + 1 \qquad \text{dividing by } (a - b)$$
$$b = b + 1 \qquad \text{substituting since } a = b$$
$$0 = 1 \qquad \text{subtracting } b \text{ from both sides.}$$

If $0 \neq 1$, then this work must contain a mistake. An important part of mathematics is deducing where mistakes happen. If you want to find the misstep on your own, stop reading and tinker with the steps to find our illogical move.

If you are ready for the move that led us to claiming $0 = 1$, find the line $a(a - b) = (b + 1)(a - b)$. In that line, we divide both sides by $(a - b)$ to get the next line that states $a = (b + 1)$. Easy enough — indeed, easy enough to overlook what just happened. Remember from the first line of the proof that $a = b$. Therefore, $a - b = 0$ and saying $a(a - b) = (b + 1)(a - b)$ simply states that $0 = 0$. More importantly, a and $b + 1$ have nothing to do with the equality since the presence of $(a - b)$ on both sides inherently forces our equality of $0 = 0$. When we divide both sides by $(a - b)$, we implicitly assume that $(a - b)$ does not equal 0 as it can lead to the type of contradiction we just saw.

In our proof, our misstep of dividing by zero led to an erroneous conclusion that $0 = 1$. Mathematics often extends results. Let's see this here. We can pick any x and y such that $x \neq y$ and conclude that x and y are equal. The key again will be dividing by zero. First, let $d = x - y$, which gives us $x = d + y$. We begin, as before, with $a = b$.

$$a = b \qquad \text{our initial identity}$$
$$a(a - b - d) = b(a - b - d) \qquad \text{multiplying both sides by } (a - b - d)$$

$$a^2 - ab - da = ab - b^2 - db \qquad \text{distribution}$$
$$a^2 - ab = ab + da - b^2 - db \qquad \text{adding } da \text{ to both sides}$$
$$a(a - b) = a(b + d) - b(b + d) \qquad \text{factoring both sides}$$
$$a(a - b) = (b + d)(a - b) \qquad \text{another step of factoring}$$
$$a = b + d \qquad \text{dividing by } (a - b)$$
$$a + y = b + d + y \qquad \text{adding } y \text{ to both sides}$$
$$a + y = b + x \qquad \text{substituting } x = d + y$$
$$b + y = b + x \qquad \text{substituting } a = b$$
$$y = x \qquad \text{subtracting } b \text{ from both sides.}$$

So, pick your x and y. Take $x = \pi$ and $y = 3.2$. Then, our "proof" concludes $\pi = 3.2$, which isn't even the proper way to round π.[a] You could also let $x = 0.\overline{9}$ and $y = 1$. Then, we need $d = 1 - 0.\overline{9}$. What is d? Said another way, what is the distance between $0.\overline{9}$ and 1? It isn't 0.00001 or 0.000000000001. In fact, it's precisely 0. There is no difference between $0.\overline{9}$ and 1 other than the way in which we write that number. This can seem mysterious and even suspect. Yet, it's actually something we already know. Consider $1 = 1/3 + 2/3 = 0.\overline{3} + 0.\overline{6} = 0.\overline{9}$. So, even the seemingly impossible can be possible. But we also see that sometimes, a misstep can make the impossible seem possible.

1.5. Baseball's 100 Year Error

Having $0 = 1$ is connected to an error in baseball. It turns out that every baseball game, ever played, has violated a rule of the game, unless, of course, zero does, indeed, equal one. Baseball has had a mathematical error in its rule book for over 100 years [42]. The home plate has the dimensions shown in Figure 1.4(a). As such, the plate can be seen as combining a rectangle and a isosceles right triangle as seen in Figure 1.4(b). Let's turn our attention to the gray right triangle in Figure 1.4(b). It's both legs have length 12. By the Pythagorean theorem, this right triangle satisfies $12^2 + 12^2 = 17^2$ or $144 + 144 = 289$, which would mean $288 = 289$. If we subtract 288 from both sides of the equation, we get $0 = 1$. Said another way, every home plate in every baseball game has violated the rules of baseball, unless, of course $0 = 1$ since that would allow $288 = 289$.

---

[a]An erroneous "proof" led to Indiana almost passing a bill in 1897 declaring that $\pi = 3.2$.

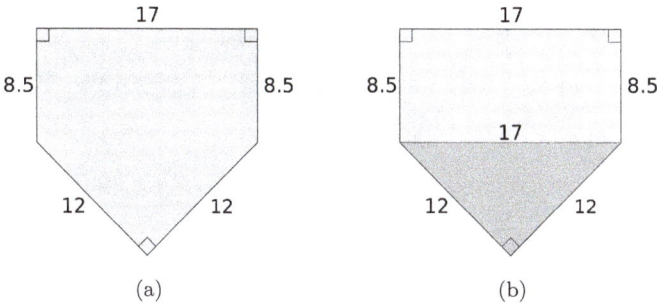

Figure 1.4: The specifications for the home plate are geometrically impossible!

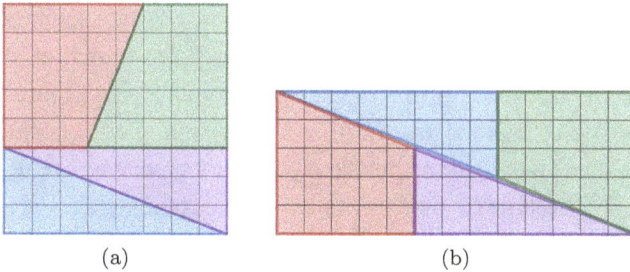

Figure 1.5: Can 64 possibly equal 65? It does if these puzzle arrangements are possible.

How far is the home plate from a 90 degree angle? We have a triangle with sides 12, 12, and 17. From the Law of Cosines, we can find the angle of interest from

$$\cos(C) = \frac{12^2 + 12^2 - 17^2}{2(12)^2} = -\frac{1}{288}.$$

So, the home plate has an angle of 90.1989 degrees. It's off by just under 0.2 of a degree. If you allow baseball to round, then the home plate satisfies the rule!

1.6. The 64 = 65 Fib

Tomfoolery need not only be with equations. Consider the 8 × 8 square in Figure 1.5(a). The area is clearly 8 × 8 = 64. Now, let's rearrange the pieces to get the rectangle in Figure 1.5(b). Note that this rectangle is 5 × 13, having an area of 65. So, it appears that we just produced a geometric argument

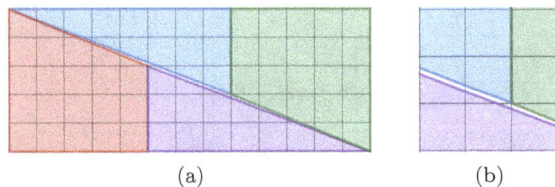

Figure 1.6: On closer inspection, $64 = 65 - 1$.

that $64 = 65$, or subtracting 64 from each side, again that $0 = 1$. How can this be true? Does this puzzle magically violate the conservation of mass? Sometimes, the best way to uncover an error is to look more closely. It turns out that the pieces don't quite fit in Figure 1.5(b), which can be hard to perceive given that I intentionally used thick lines to outline the pieces. If I reduce the thickness of the lines, I get the puzzle in Figure 1.6(a). In Figure 1.6(b), I've zoomed into the hole in the puzzle. Can you anticipate the area of the space between the puzzle pieces? It's precisely 1 unit, giving us $64 = 65 - 1$. This type of puzzle is called a dissection puzzle and is credited to Sam Loyd, who was a prolific puzzle maker; his book *Cyclopedia of 5000 Puzzles* was published in 1914 by his son with a recent version published in 2007 [43].

We saw that an error led to a claim that $0 = 1$. How is our puzzle concluding that $64 = 65$? The answer is connected to the Fibonacci sequence, which is $\{0, 1, 1, 2, 3, 5, 8, 13, 21, \ldots\}$. We need to remember only the first two elements of the sequence, 0 and 1. Each successive element in the sequence equals the sum of the previous two elements. Let $F_0 = 0$ and $F_1 = 1$. Then, $F_n = F_{n-1} + F_{n-2}$. So, $3 = 2 + 1$, $5 = 3 + 2$ and $8 = 5 + 3$. The key to the puzzle is that the fraction F_{n+1}/F_n becomes closer and closer to the golden ratio, denoted $\varphi \approx 1.618$, as n increases. For example, $21/13 = 1.615$.

Since F_{n+1}/F_n is converging to φ, $F_{n-1}/F_n \approx F_{n+1}/F_{n+2}$. Cross-multiplying, we get $F_{n-1}/F_{n+1} \approx F_n/F_{n+2} \approx 0.381966011$, which will help us to demystify our puzzle. The pieces fit perfectly in Figure 1.5(a). In Figure 1.5(b), they almost fit. Let's focus on the diagonal of the rectangle in Figure 1.5(b), which is formed by a right trapezoid and a right triangle. The key is that the trapezoids and triangles don't have the same

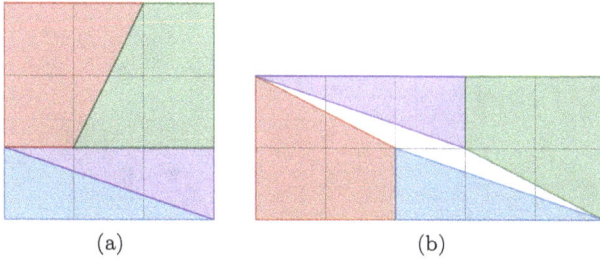

Figure 1.7: The Fibonacci sequence helps to construct a smaller version of this section's puzzle.

slope, resulting in a bent diagonal. The trapezoid has a slope of $2/5 = 0.4$. The triangle has a slope of $3/8 = 0.375$. Note that the numbers in these slopes are terms in the Fibonacci sequence $2, 3, 5, 8$. Our property $F_{n-1}/F_{n+1} \approx F_n/F_{n+2} \approx 0.381966011$ helps us to know that these slopes will be close in value.

This gives us insight into making similar puzzles. Let's look at the pieces in Figure 1.5. The lengths of the bases of the right trapezoids are 3 and 5. The legs of the right triangles are 3 and $3 + 5 = 8$.

First, let's make a smaller puzzle. We'll use the Fibonacci numbers 1, 2, 3, and 5. Our right trapezoids will have bases of length 1 and 2. Our right triangles will have legs of length 1 and $3 = 1 + 2$. We fit these together as a square as seen in Figure 1.7(a). We reform them into a rectangle (with the missing area) as seen in Figure 1.7(b). The square has area $3 \times 3 = 9$ and the rectangle, including the missing portion, has area $2 \times 5 = 10$. The missing portion is very clear since 10% of the rectangle's area is missing!

So, let's make a larger puzzle using the Fibonacci numbers 5, 8, 13, and 21. Our right trapezoids will have bases of length 5 and 8. Our right triangles will have legs of length 5 and $13 = 5 + 8$. We fit these together as a square as seen in Figure 1.8(a). Moving the pieces forms a rectangle (with missing area) as seen in Figure 1.8(b). The square has area $13 \times 13 = 169$ and the rectangle, including the missing portion, has area $8 \times 21 = 168$. This time, the missing area is difficult to see as it constitutes less than 0.6% of the rectangle's area.

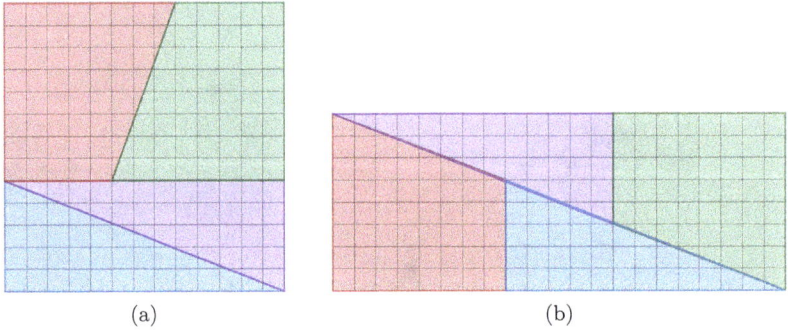

(a)   (b)

Figure 1.8:   The Fibonacci sequence can lead to a larger version of this section's puzzle.

Numbers play an inherent role in sports. They keep the score and also, if tracked with insightful statistics, can enable us to evaluate the performance of a team or individual. Still, we must guard against seeing sports analytics as truth. Most analytics derive from mathematical models and have their limits. They can even lead to puzzling results, which is the topic of the next chapter.

Chapter 2

Trick Plays

Math should be easy enough. Do the math and find the answer. With "data", you are looking for insight more than an answer. You might even find confusing insights that almost seem like magic tricks. In this light, look at the two images in Figure 2.1. Which image has a longer vertical line segment? Don't break out a ruler. What do you see with your eyes?

The vertical line segments have the same length even though one looks longer to many of us. Just because you see something doesn't mean, no matter how certain you are that you see it, that it is real.

Psychologists John van der Kamp and Rich Masters examined whether goalkeepers can influence penalty-takers' behavior by mimicking the illusion in Figure 2.1 [29, 40]. A soccer goal is eight yards wide and eight feet high. Penalty shots are kicked at an average speed of 70 miles per hour from only 12 yards from the goal line. At this speed, it only takes about 0.35 seconds for the ball to leave the kicker's foot and arrive at the goal line. Note that the blink of an eye has been measured at 0.4 seconds. With so little time to react, a goalie must anticipate where the ball will be kicked.

In the study, the goalie assumed the postures in Figure 2.2, which from left to right were labeled arms-out, arms-up, arms-down, and arms-parallel. A goalie standing with the arms-out or arms-up posture mimicked the illusion on the left in Figure 2.1, which generally appears to have a longer vertical line. More importantly, when the goalie's arms were positioned with arms-down or arms-parallel, the goalie was perceived as being smaller, and

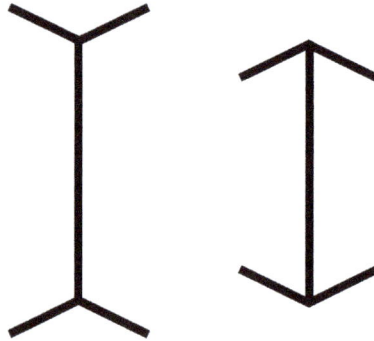

Figure 2.1: Which vertical line is longer?

Figure 2.2: Postures of a goalie from a study by van der Kamp and Masters: (from left to right) arms-out, arms-up, arms-down, and arms-parallel.

the ball was directed closer to the goalkeeper, which is more desirable for the goalie. As such, van der Kamp and Masters' study established that a goalie's posture affected how tall the goalie appeared, which influenced the accuracy of a penalty kick.

Do you want to improve your golf game? For this, we turn to another illusion. If you concentrate only on the black hole in Figure 2.3(a), it will look smaller than when you concentrate only on the black hole in (b). Psychologist Jessica Witt of Purdue University found that this illusion could affect your putting. Witt's research group used the optical illusion in Figure 2.3 [66,67]. The configuration in (a), with the larger-looking golf hole in the center, led to a 10% improvement in putting. Witt's other work has shown that softball and tennis players who are hitting well think that the balls look bigger.

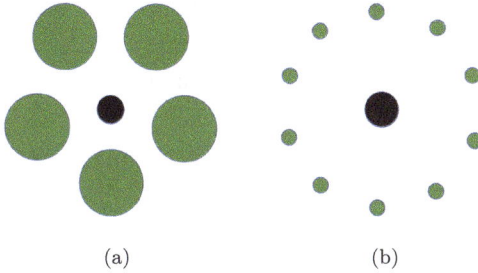

Figure 2.3: Some holey images that can play with our perceptions.

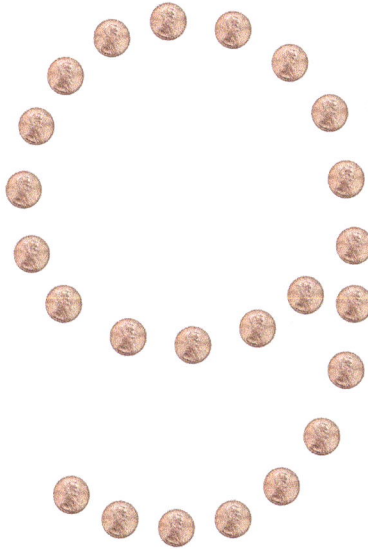

Figure 2.4: A collection of pennies to count.

2.1. Mathematical Magic

Surprises can also come from simple counting. Let's use a handful of pennies to illustrate this.

To begin with, I'll place a group of pennies in the configuration of the number 9, as seen in Figure 2.4. Follow these steps:

1. Pick a double-digit number less than 20.
2. Starting at the tail of the 9, count up and around the nine counter-clockwise until you reach your number. Keep track of where you land.

3. Beginning at the last coin touched and counting it as 1, continue to count clockwise around the circle (not going down the tail this time) until you again reach your selected number. Point at that penny.

4. Here's the magic — I've already predicted the penny at which you are pointing. Look at Figure 2.5 to see my prediction. It's the highlighted penny.

Think about how the trick works and see if you can uncover the secret. How did I know in advance, regardless of the number you picked in the first step for the trick, which penny would be your last one when the trick ends? Hint — the magic is quite a tail! (Pun is entirely intended.) Once you have got it or are ready with the answer, turn the page.

This trick was created by Martin Gardner, who is said to have brought "more mathematics, to more millions, than anyone else." He presented the trick in his 1956 book *Mathematics, Magic and Mystery* [32]. Why do math? Because you can amaze your friends!

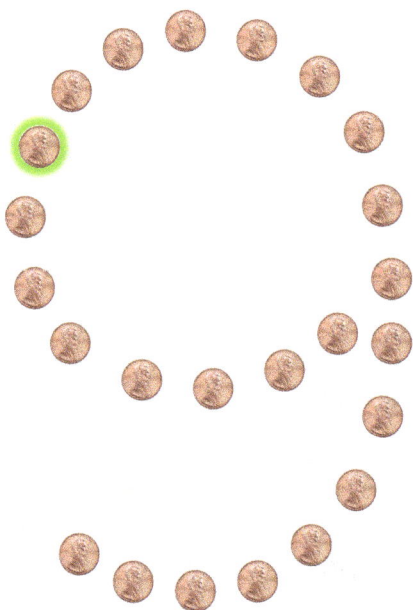

Figure 2.5: A mathematical prediction.

Did you figure out the secret of the magic? Let's walk through the trick. You pick a number and count up the tail and then counter-clockwise around the circular part of the nine. Then, you reverse the direction. When you get to the highlighted penny in the configuration to the left below, how many pennies would you have left in the tail? You'll always have the numbers of pennies in the tail left to count. In this case, you'll have eight more pennies to count. This gives us the prediction we saw earlier and pictured in the configuration to the right below.

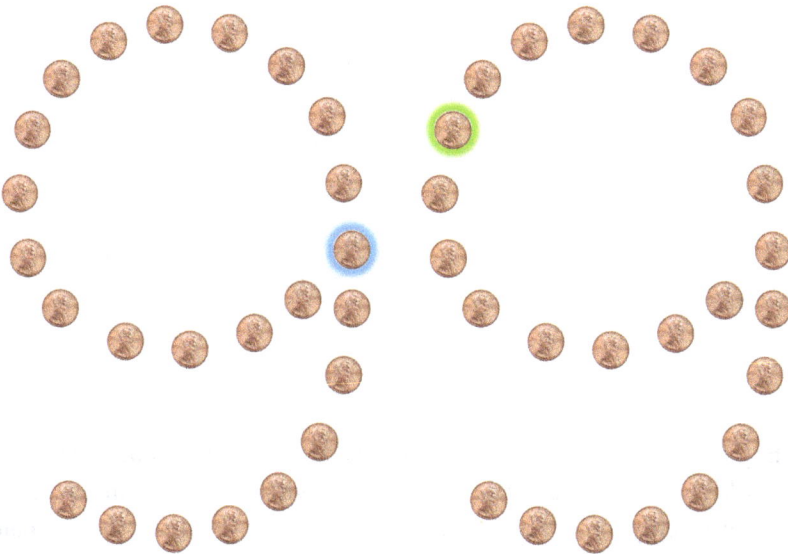

You can use whatever change you have in your pocket as long as you have enough to create a 9. The tail can have 5, 6 or 13 coins!

Let's try the trick again with a different number of coins as seen in Figure 2.6 to ensure you are ready to do this magic on your own. We will use the confirmation in Figure 2.6. To begin with, note the number of pennies in the tail of the 9. Use this number to create your prediction of where you will be pointing when the trick ends. Let's check your prediction by following our magical steps:

1. Pick a number between 8 and 15.
2. Starting at the tail of the 9, count up and around the nine counter-clockwise until you reach your number. Keep track of where you land.

Figure 2.6:   Another magical layout of pennies.

3. Beginning at the last coin touched and counting it as 1, continue to count clockwise around the circle (not going down the tail this time) until you again reach your selected number. Point at that penny. It should match your prediction.

If your prediction didn't match where you landed, you could have miscounted, or it may help us to read the following explanation of the prediction, which is on the next page.

With this new configuration of pennies, we have only five pennies in the tail. When you are counting clockwise around the circle and land at the highlighted penny in the configuration on the left in the following figure, you will have five pennies left to count. Counting those five pennies, lands us with our prediction seen in the configuration on the right in the following figure.

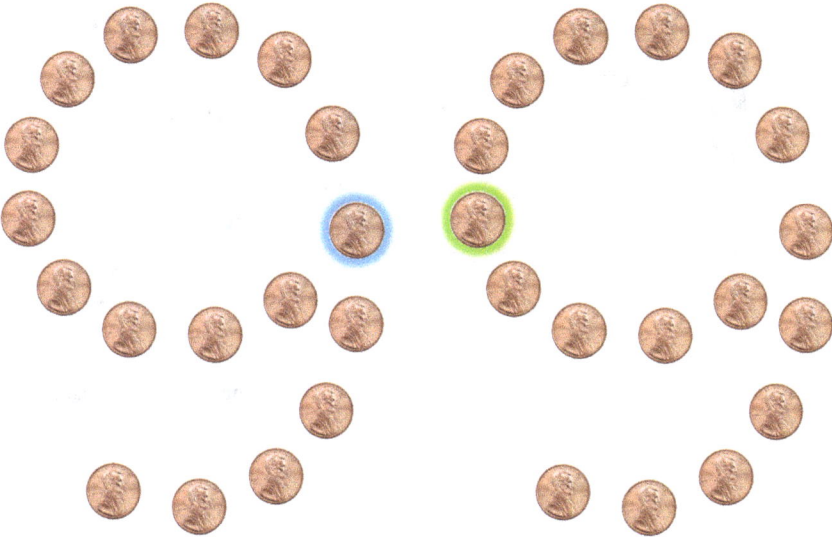

Now, you should be ready to break out the coins in your pocket or even search under the couch cushions for coins to share some mathematical magic.

2.2. Paradoxical Insight

How tricky can the actual data of sports be? Let's look at a puzzle I created for the Numberplay section of *The New York Times* reported by Gary Antonick [2]. In Figure 2.7, we see the shot charts of two players at Davidson College's basketball court, which is where I teach and lead my sports analytics group. The player in (a) shot very well within the three-point arc at a rate of 62%. Such shooting is often called hot shooting, which is why the region is colored red. Player (b) shot 54.5% in that region. Player (a) also shot three-pointers at a higher rate of 35% versus player (b)'s 25%. Shooting at 25% is often called cold shooting, leading to the blue coloring. If you combine all the shot attempts together, you get an overall shooting percentage. One player had an overall shooting percentage of 50.3% and the other had 48%. Surprisingly,

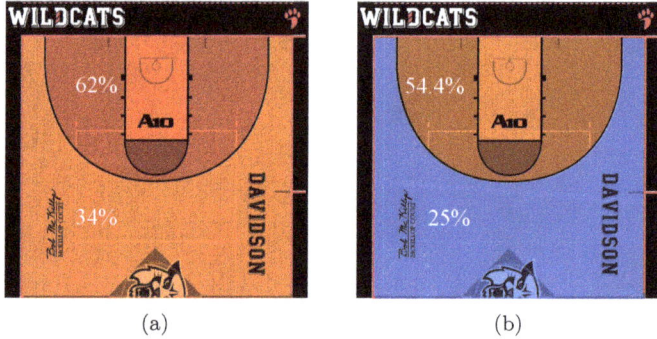

(a)   (b)

Figure 2.7:   Shot charts with overall shooting percentages that may be surprising.

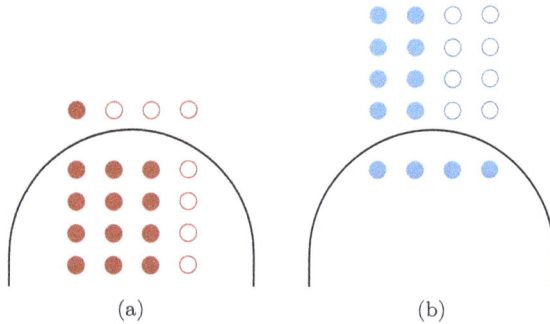

(a)   (b)

Figure 2.8:   A small example of Simpson's paradox in basketball. Circles represent shots with filled circles representing scoring attempts.

the player with a lower overall shooting percentage is player (a). How could this be? Let's look at an example to get some basic intuition.

This is an example of Simpson's paradox. The paradoxical behavior lies in how the data is displayed. By not disclosing the number of shots each player took, it can appear that the numbers don't add up.

To see how this can happen, look at Figure 2.8. Circles represent shots with filled circles representing scoring attempts. The red player in Figure 2.8(a) made 1 of 4 shots from behind the 3-point line and 12 of 16 from within the arc. So, the red player shot 25% for three-pointers and 75% for two-pointers but 13 of 20 or 65% overall. Now, compare this to the blue player in Figure 2.8(b) who shot 100% for two-pointers and 50% for

Table 2.1: Comparing the shooting of James Harden and Dirk Nowitzki in the 2017–2018 season.

Player	FG%	3pt%	FT%	Total%
James Harden	44.9% (651/1449)	36.7% (265/722)	85.8% (624/727)	61.9%
Dirk Nowitzki	45.6% (346/758)	40.9% (138/337)	89.8% (97/108)	57.5%

three-pointers but 12/20 or 60% overall. The unbalanced number of shots in the regions leads to the unintuitive result when you see only the overall percentages.

Let's see the underlying numbers that generated the example in Figure 2.7. Note that other numbers can be derived to create the same result. The example had player 1 making 31 of 50 two-point shots and 17 of 50 three-point shots. So, this player made 48% of her total shots. Player 2 made 68 of 125 two-point shots and only 5 of 20 three-point shots, resulting in her making 73 of 145 shots.

We have a paradox but I made this example up. How real is this phenomenon? Can we expect it in actual data we'd find in sports or our world? Let's look at the 2017–2018 season in the NBA. We'll compare James Harden to Dirk Nowitzki on field-goal, three-point and free throw shooting percentages as seen in Table 2.1. We see that Nowitzki has a better shooting percentage in all three categories. However, from the last column, when we combine the shots, Harden is over 4% better in shooting than Nowitzki. Again, the key is to look at the frequency of each type of shot and their associative percentage. Harden had almost seven times as many free throw attempts as Nowitzki and made over six times as many. In his final shooting percentage, Harden's 727 free throws, of which he made 85.8%, constitutes 25% of his overall shots. For Nowitzki, 108 free throws are only about 9% of his total shots. This is, in part, what pulls Harden's overall shooting percentages over Nowitzki's. Said another way, a key to finding this paradox for Harden and Nowitzki is knowing that Harden has a large number of free throws relative to his other shots and he has a high shooting percentage for free throws.

Let's see another in sports. Comparing pass completion of Jameis Winston and Jared Goff, we see Winston having the advantage during wins

Table 2.2:   Comparing pass completion of Jameis Winston and Jared Goff in the 2018 NFL season.

Player	During a win	During a loss	Total
Jameis Winston	67.50% (81/120)	63.18% (163/258)	64.55% (244/378)
Jared Goff	66.27% (281/424)	60.58% (83/137)	64.88% (364/561)

Table 2.3:   Data of Alaska and America West Airlines from 1991.

	Alaska			America West		
	On time	Delayed	Delay%	On time	Delayed	Delay%
Los Angeles	497	62	11.1	694	117	14.4
Phoenix	221	12	5.2	4840	415	7.9
San Diego	212	20	8.6	383	65	14.5
San Francisco	503	102	16.9	320	129	28.7
Seattle	1841	305	14.2	201	61	23.3
Total	3274	501	13.3	6438	787	10.9

and losses for pass completion in Table 2.2. A key in finding this example of Simpson's paradox is seeing the difference in the number of games the teams won. The Buccaneers won only 3 of the 9 games Winston started. The Rams won 13 of the 16 games Goff started in the regular season. Both players have higher pass completion during winning games and Goff's 424 attempts are over 75% of his total attempts in the regular season. Winston's 120 attempts are just under 32% of his total attempts. Again, this difference leads to Goff having a higher overall pass completion even though it is lower during wins and also during losses.

This paradox plays important roles outside sports. Let's dig into some numbers, specifically transportation data [45]. Airlines work on reliability, so passengers reach their connections and destinations on time. Look at the data in Table 2.3 for Alaska Airlines and America West Airlines from 1991 for five cities on the west coast of the United States. That's quite a few numbers, so let's focus on the last row of the table. Alaska has 13.3% of its flights delayed while America West has only 10.9%. Easy enough; book America West! Before securing that itinerary let's look city by city. Alaska has 11.1% of its flights delayed into Los Angeles whereas America West has 14.4%. In fact, Alaska Airlines has a lower percentage of delayed flights than America West for every city. Easy enough; book Alaska! Wait, what? The apparent conundrum stems again from Simpson's paradox. Alaska and America West airlines could both note a better record in 1991. It all depends on the view of the data. If you are presented with such data, you want to be sure to get a table like Table 2.3 so you get a full view!

Analytics can be tricky. When we see something paradoxical or surprising, an important lesson of this section is to dig deeper. When you look at data in a variety of ways, what was once surprising can become explainable.

2.3. Disappearing Dissection

An important mathematical step made the previous paradoxes explainable. We looked at the data in a different way. Mathematics can make the magical logical. Let's see another puzzle that again underscores how much tricky counting can be. This also appeared in *The New York Times'* Numberplay reported by Gary Antonick [2]. In Figure 2.9(a), we have a group of 14 basketball players.

If you cut out the puzzle along the straight black lines and interchange the left and right pieces on the top row, then you get the configuration in Figure 2.9(b). How many players do you count now? Count carefully.

How did this happen? It can seem impossible. When you are ready for the answer, look at the end of the section. To make it harder to accidentally

(a)   (b)

Figure 2.9:   A group of 14 basketball players in a puzzle (a) containing three pieces. Swapping the top two pieces in (a) creates the configuration of the puzzle in (b). How many players do we have now? Cartoon by Boluwatife Odumosu.

see what's happening, the answer is at the end of the chapter with a necessary spoiler alert so you can keep from reading the answer until you are ready.

2.4. Footnotably Perfect

We must be careful about sports statistics, else they will seem tricky. Let's look at an example from Major League Baseball. Ty Cobb played his last game on September 11, 1928. He'd hit 36.7% of the time he was at the plate, corresponding to a 0.367 batting average, which would be the leading career batting average for a century. Actually, a teammate of Cobb had a perfect career batting average, hitting every time he was at the plate. Indeed, there has been a 1.000 batting average. This impressive batting average was, in fact, partially due to Ty Cobb.

In May 1912, Ty Cobb's notorious temper erupted as he lept into the stands and cleat-stomped a New York fan who heckled Cobb. The act led to a ten day suspension. Cobb's Tiger teammates struck in protest of his suspension, hoping to underscore their hope that the league needed to protect its players from grossly insulting fans. Rather than forfeit their next game against the Philadelphia Athletics, the Tigers filled their roster with local Philadelphia semi-pro and amateur players. Note that it wasn't the forfeited loss to the Athletics that likely motivated Tiger management to use replacement players. Such a forfeit could also lead to a $5,000 fine or a loss of the franchise. In fact, in 1902, lacking the minimum number of players required to compete, the Baltimore Orioles forfeited their game against the St. Louis Browns. This led to a loss of the franchise and a change of ownership. So, on May 18, 1912, the replacement players took the field. Among them was Ed Irwin, playing in his only big league game, who walked to the plate three times. He hit two triples, giving him a career batting average of 0.666, which, of course, is better than that of Ty Cobb.

Irwin doesn't have the best career batting average in MLB. There are perfect career batting averages. John Paciorek played the last game of the 1963 season as a Houston Colt of the .45s. He stood in the batter's box five times leading to three hits and two walks. The game ended when the visiting Mets went scoreless in the top of the 9th inning and lost 13 to 4. With that, Paciorek finished the season and his Major League career with a perfect 1.000 batting average. A back injury would move him to the Minor League where he didn't play during the 1964 season but played in the minors for the following five seasons [61].

Although interesting, Paciorek and Irwin's batting averages are more footnotable than statistically significant. In fact, the small number of bats is what made such feats possible. Paciorek isn't the only player who has batted perfectly on his first day. A dozen "four hit wonders" have hit 4 for 4 in their debuts. In September 1912, Casey Stengel debuted for the Brooklyn Dodgers in a game against the Pittsburgh Pirates. Along with his hits, Stengel batted in two runs and stole two bases in the Dodgers 7-3 win. In 1933, Russ Van Atta walked to the mound to pitch for the Yankees in his debut game. When the Senators left the field without scoring in the ninth inning, Van Atta had recorded four hits in four times at bat and pitched a 17-0 shutout.

There is only about a 0.4% chance of going 4 for 4 at the plate with a batting average of 0.250 since $0.25^4 \approx 0.004$. However, over 15,000 players have played Major League Baseball. A quick calculation indicates that we could have expected about $(0.004)(15000) = 60$ players to have gone 4 for 4 on the opening day. So, we actually see less than the expected number of players achieving this feat, at least as indicated by our quick calculation. This underscores how there is more to the analytics than the number. Our calculation is quick, often indicating it could overlook factors. Further, do we really expect hitters to bat as well on their first days as when they've spent time in the league? Nerves simply must be a factor when one walks, for the first time, to the home plate in a Major League stadium.

This underscores an important aspect of sports analytics. We must also be careful not to lean too heavily on our computations. We must be mindful of context, which in this section has often been sample size. While it is true that John Paciorek had a 1.000 batting average, the leaderboard of batting averages for Major League Baseball will have a threshold for the number of times a player must have been at the plate to make the list. This keeps such spurious results from entering our results.

Keep such "trick plays" in mind. Surprises can mean that we have uncovered new insights. It can mean that we've overlooked an aspect of the game, like achieving a career batting average of 1.000 from standing at the plate a total of two times in one's MLB career. When surprises arise in data analytics, there is a second stage of determining why.

Figure 2.10:   If we reorder the 14 basketball players, how we pick up a 15th becomes clear.

Spoiler Alert

So, let's see why our illusion in Figure 2.9 worked. Recall that in Figure 2.9(a), we had 14 basketball players. We swapped the position of two pieces of the puzzle and suddenly had 15 players. A key to uncovering the trick of this magic is reordering the basketball players. We line up the players as seen in Figure 2.10(a). This puzzle has two pieces, one on top and one along the bottom. Shifting the top piece to the left easily produces 15 players. If you look carefully, the players in Figure 2.9(a) match the players in Figure 2.10(a). Similarly, the players in Figure 2.9(b) match the players in Figure 2.10(b). So, swapping the pieces on the top row of the puzzle in Figure 2.9(a) has the same effect as shifting the top piece of the puzzle in Figure 2.10(a). This is an important mathematical lesson. Things can be obscured until we look at them in a different arrangement.

Chapter 3

Finding Hall of Famers

When you first receive data, you need to get to know it. In 1977, John Tukey published a highly influential book entitled *Exploratory Data Analysis*. He recommended starting analysis by graphing the data.

There are various aspects of graphical data that can lead to insights. Let's try a common children's puzzle requiring spotting the difference. In Figure 3.1, we see a group of images of a Babe Ruth baseball card. Which picture is different from the others? Keep looking as the difference only emerges when you spot the variation. With data, there may be a difference but it is likely you won't know it's there until you see it. In the case of Figure 3.1, you knew when you looked at the picture that a difference exists.

We have data. How do we jump into it? Graph it. Then, look for similarities and differences. Similarities can uncover patterns. Differences can signal outliers. In sports, those outliers may be future Hall of Famers!

3.1. Finding the Oddball

So, let's dive into some data detailing the number of home runs for American League teams in 1920. Boston had 22 home runs, Chicago had 37 and so forth, producing the graph in Figure 3.2(a). Note the large number of home runs hit by the New York Yankees. We are seeing a difference and in this case an outlier. In home run production, the Yankees more than doubled any other team in the American League. In particular, note how quickly we spotted the outlier in the graph, which is why Tukey emphasized graphing as a means of familiarizing oneself with data. Even so, we don't see why it's true. We just know that there is a difference.

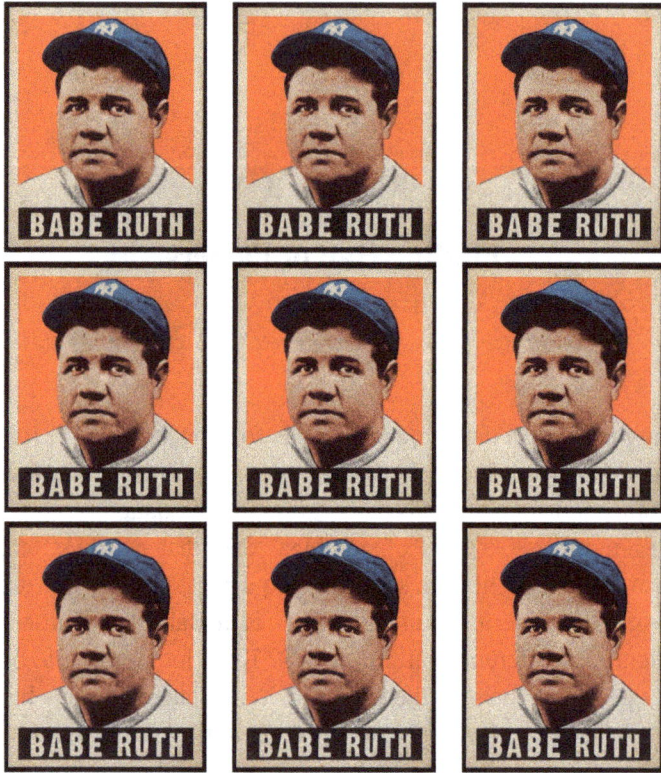

Figure 3.1: Which Babe Ruth baseball card isn't like the rest? Spotting differences can be a great way to start with data.

So, we will look at the data in a different way. Now, rather than examining the home runs by eight teams, we'll look at the number of home runs hit by every player in the American League in 1920. That's over 250 in number. You can page through over 250 player home run statistics or graph it. Figure 3.2(b) contains a graph of the number of home runs by every American League player who hit at least one home run in 1920. Do you see the outlier? If we dig deeper, we learn that the outlier is Babe Ruth, the original home run king. He smacked 54 homers, which is more home runs than any other American League team that season. In fact, only the National League's Philadelphia Phillies managed to hit more home runs than Ruth. Ruth hit 54 and the Phillies' entire team hit 64.

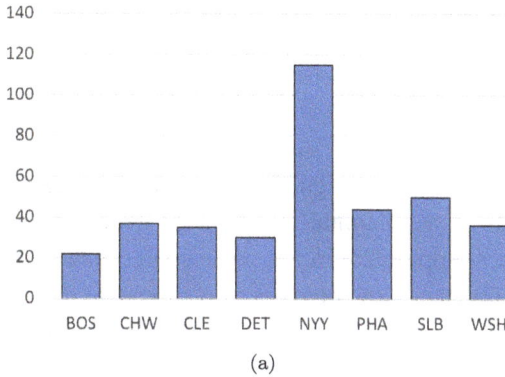

(a)

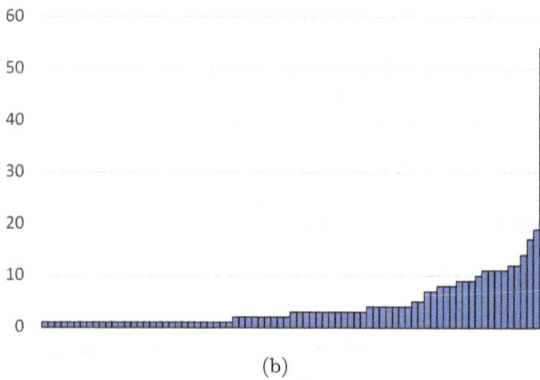

(b)

Figure 3.2:   Number of home runs in the American League of Major League Baseball in 1920.

Note how we looked at the data by team and then by the players to gain our insight. This is often true with analytics. You have your data and often need to transform the dataset. You visualize and then you analyze and interpret. Keep in mind that sometimes, this will lead to a need to further work with the data, visualize again and analyze your new results. For example, we can add the date on which each home run was hit. Then, we might ask if the home runs were hit at the same pace over the season. Or else, we could examine if more home runs were hit at home or on the road. Soon, you can have many questions. Mathematical work can be a cycle. In analytics, you may receive data; you need to transform it, and then visualize it. At that point, you analyze and interpret the visualization and decide if you need to transform or augment the dataset for further exploration.

3.2. Spatial Difference

Tukey encouraged looking at data to get used to it. Our data wasn't all that large as even 250 numbers isn't all that many. However, we are now in the era of big data. In sports, the NBA began logging enormous datasets in the fall of 2013 when the league signed a contract with STATS LLC to place SportVU cameras in every arena. The cameras were generally perched in a stadium's rafters, snapping 25 frames per second. The images were fed into a computer that translated the pictures into a treasure trove of data.

SportVU began in 2005 when Israeli missile defense specialist Miky Tamir decided to explore the use of defense system optical tracking technology to create sports stats. Before long, such methods of computer vision were tracking soccer players, the ball, and referees in arenas. In 2008, STATS purchased the SportVU technology and were soon adapting it to basketball. Among various differences between the sports, the smaller space on a basketball court leads to more congestion and the need for multiple angles to track jersey numbers in order to identify players. So, STATS used six cameras to track basketball and three for soccer [20].

What exactly did the SportVU cameras produce? For each game, the NBA received an XML file containing raw coordinate data for each player and referees along with the UNIX time code and game clock. Coordinate data was also supplied for the ball — with this being 3D data. This data was supplied for each 1/25 of a second of the game, producing a 40–45 megabyte file for each game. Teams received the data on every game — everything except the referee data, which is analyzed only by the NBA.

Take a moment and consider the amount of data and possible insight in such a file. One can easily track how far any given player (or the ball for that matter) traveled in the duration of a game. For example, Bradley Beal of the Washington Wizards led the league in 2018–2019 for distance traveled per game at a total of 2.75 miles per game. We can also quickly compute velocities or acceleration of players, referees, or the ball. For example, Buddy Hield of the Sacramento Kings led the league (among the players who played at least 40 games) with an average speed of 4.71 miles per hour.

Among the many pieces of data available from this dataset, we know the shot location of every shot. In fact, the NBA has gone back and recorded the location of every shot of every NBA game back into the 1990s. We can use this to see the evolution of how NBA championship teams play. Let's look

(a)

(b)

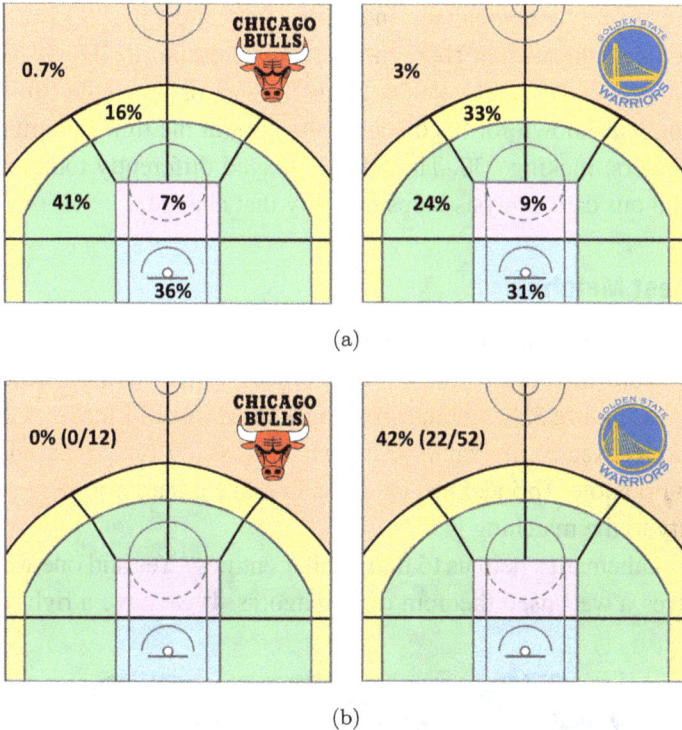

Figure 3.3: Shots during the NBA playoffs of the '97–98 Chicago Bulls and the '17–18 Golden State Warriors.

at the change of distance shooting. To aid in this, we'll break the court into five regions, colored orange, yellow, green, purple and blue in Figure 3.3. The three-point line is at the border of the green and yellow regions. So, the orange region is a very distant shot. Let's compare shots taken over the playoffs by two championship teams: '97–98 Chicago Bulls and '17–18 Golden State Warriors.

First, we compare the percentage of shots taken in each region. We see the differences in Figure 3.3(a). Note how the Bulls are shooting very long-range shots less than 1% of the time and the Warriors select the long-range shot 3% of the time. Even more telling are the shooting percentages from these distances. The Bulls did not make any of their 12 shots from that region. The Warriors made 22 of their 52 attempts, shooting 42% from a very long-range. We see this graphically in Figure 3.3(b).

You can see the comparison in another way. Take the '92 Bulls. They attempted 454 three-point shots in the regular season. In '18–19, Stephen Curry attempted 810, making 354. In his five seasons with the Bulls, Steve Kerr, who was known for his distance shooting at the time, attempted 898 three-pointers, making 430. The NBA is played differently today than the 1990s and our data analysis helps quantify that change.

3.3. Best Matches

Let's play another matching game. In Figure 3.1, we looked at images of Babe Ruth in order to locate a difference. This time, we'll look for similarities. Figure 3.4 contains six images of Abraham Lincoln. Only two of them are exactly the same. Can you find which two? The answer is given in Footnote[a] (printed upside down to make it harder to read given its proximity to the matching game).

Can mathematics help us to find similar entities? Yes and one way to do so involves a well-used theorem of Pythagoras. If we have a right triangle

(a)   (b)   (c)

(d)   (e)   (f)

Figure 3.4: Which two images of Abraham Lincoln are the same?

---

[a] The images in Figure 3.4(a) and (e) are the same.

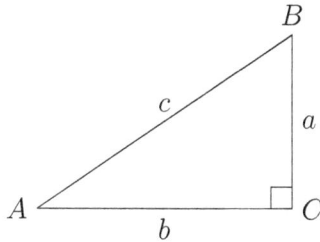

Figure 3.5: A right triangle.

as seen in Figure 3.5 with sides of length a, b and c, then the Pythagorean theorem states that $c^2 = a^2 + b^2$.

This theorem is useful in finding the distance between two points. For instance, the distance between $(2, 7)$ and $(5, 11)$ is $\sqrt{(5-2)^2 + (11-7)^2} = \sqrt{9+16} = \sqrt{25} = 5$. This can also be extended to 3D to find the distance between two points such as $(2, 7, 4)$ and $(5, 11, 6)$ which equals $\sqrt{(5-2)^2 + (11-7)^2 + (6-4)^2} = \sqrt{9+16+4} = \sqrt{29}$. In a similar way, we can blast into 4D by extending this theorem further. We'll begin outside of sports and then see this idea within athletics.

3.4. Finding Closest Musicals

Let's move into the fourth dimension to compare taste in some Broadway musicals. Suppose Tony, Emmy and Graham rate the musicals *The Lion King*, *Wicked*, *The Phantom of the Opera*, and *Hamilton* between -5 and 5. A rating of 5 correlates to wanting to see the musical (for the first time or again) and -5 means definitely not wanting to see it again. To keep things general, I'll let a random number generator create ratings for the musicals. The ratings appear in Table 3.1.

We treat each person's ratings as a point in the fourth dimension. So, Tony's ratings becomes the point $(0, 5, 3, 5)$, Emmy's $(2, -4, 4, 5)$ and Graham's $(2, 3, 3, 1)$. Now, we find the distance between the points. The distance between Tony and Emmy's points is

$$\sqrt{(0-2)^2 + (5-(-4))^2 + (3-4)^2 + (5-5)^2} = \sqrt{86}.$$

The distance between Tony and Graham's points is $\sqrt{24}$ and that between Emmy and Graham's points is $\sqrt{66}$. The smallest distance we found between

Table 3.1: Broadway musical ratings for Tony, Emmy and Graham.

	Tony	Emmy	Graham
The Lion King	0	2	2
Wicked	5	−4	3
The Phantom of the Opera	3	4	3
Hamilton	5	5	1

ratings was for Tony and Graham enabling us to conclude that they have the most similar tastes in these Broadway musicals.

Table 3.1 enabled us to step into the fourth dimension. Why stop there? How about 943 dimensions? Note, that we took the columns in Table 3.1 to find people with similar tastes. We could have also taken the rows to see which musicals are similar, at least according to the ratings. Let's use that idea on a larger dataset of movie ratings. We will look at 100,000 ratings of over 900 people with the MovieLens dataset collected by the GroupLens Research Project at the University of Minnesota. The ratings are between 1 and 5 from 943 users on 1682 movies where each user rated at least 20 movies. The data was collected from September 1997 to April 1998 [33].

Let's create a point in the 943rd space containing the ratings of all 943 people for a movie. A rating of zero indicates that a movie was not rated. Every movie becomes a point. We'll pick a movie and for its corresponding point, we'll find which other movie's point is closest.

For example, consider the 1994 film *The Lion King*. The top three most similar films are *Aladdin* (1992), *Beauty and the Beast* (1991), and *Cinderella* (1950). If you were in the dataset, you could also find your ratings, turn them into a point and find the user whose ratings are closest to you. You could then see films that person liked that you haven't seen. This is one form of a recommendation system.

3.5. We are the Champions

Let's analyze NBA championship teams. In Table 3.2, you see data on championship teams for a decade in the NBA. We can use this same measure of distance to analyze the teams for similarities. To begin with, let's take

Table 3.2: Statistics for a decade of NBA Championship teams.

Year	Team	Three point (%)	Two point (%)	Free throw (%)	Total points
2018–19	Toronto Raptors	0.366	0.539	0.804	9384
2017–18	Golden State Warriors	0.391	0.56	0.815	9304
2016–17	Golden State Warriors	0.383	0.557	0.788	9503
2015–16	Cleveland Cavaliers	0.362	0.514	0.748	8555
2014–15	Golden State Warriors	0.398	0.514	0.768	9016
2013–14	San Antonio Spurs	0.397	0.517	0.785	8639
2012–13	Miami Heat	0.396	0.536	0.754	8436
2011–12	Miami Heat	0.359	0.496	0.775	6500
2010–11	Dallas Mavericks	0.365	0.516	0.777	8220
2009–10	Los Angeles Lakers	0.341	0.492	0.765	8339

only the data for three-point percentage, two-point percentage, and free throw percentage, giving us a point in 3D. If we graph all ten points, which two teams are the closest? The 2013 San Antonio Spurs are most similar to the 2014 Golden State Warriors with a distance of 0.017. We can also find the teams that are most dissimilar by these statistics. If we do for the same data, we find that the 2009 Lakers are most dissimilar to the 2017 Warriors with a distance of 0.098. Note, that we also have total points, which we'll now include in our analysis. If we look at three-point percentage, two-point percentage, free throw percentage, and total points for each team, we have a point in 4D. The two points that are closest are the 2017 Golden State Warriors and the 2018 Toronto Raptors with a distance of 80. The most dissimilar are the 2011 Miami Heat and the 2016 Golden State Warriors with a distance of 3003.

We got significantly different results when we added the Total points column. Let's look at the computation of distance for the 2017 Golden State Warriors and the 2018 Toronto Raptors. The distance would be

$$\sqrt{(0.391 - 0.366)^2 + (0.56 - 0.539)^2 + (0.815 - 0.804)^2 + (9304 - 9384)^2} \approx 80.$$

In particular, note that the difference in total points is 80, which is our final computed distance between the teams. Note that three-point percentage, two-point percentage and free throw percentage are percentages between 0 and 1. On the other hand, total points range between 6,500 and almost 10,000. As such, the difference in total points dominates our distance measure. Said another way, our final computation simply tells us which teams were most and least similar in total points. We are using four columns of data but the difference in only one leads to our indicated differences. Note that the most dissimilar teams when you use the four columns of data are the 2011 Miami Heat and 2016 Golden State Warriors, which have the biggest difference in scoring.

We didn't see this when we looked at the three columns of data for the NBA teams since they were all percentages. When we looked at the movie data, all the ratings were scaled between 1 and 5 on MovieLens. If we wanted to include total points, we need to think about rescaling the column of data so it doesn't dominate our computation. As such, using such a measure of distance requires planning, modeling decisions, and carefully mathematical thought.

Note that rather than teams, you could find similar players. You might take current and former statistics of players during college to aid in decisions of a draft. You might use players' statistics over time to discern how a career might unfold. This can help with decisions regarding trades. Such measures can help gain insight from a sea of statistics.

3.6. Image Detection

Now, let's use our mathematical measure of distance to find which images of Lincoln are the same in Figure 3.4. Before working with the images of Lincoln, let's simplify matters and use math to decide if the images in

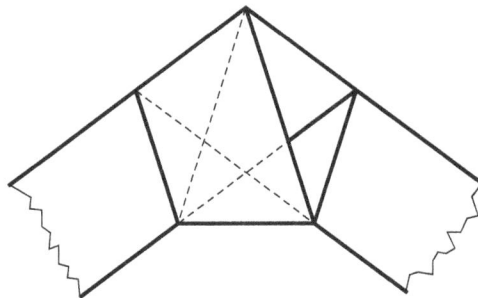

Figure 3.6:   Two 3 × 3 images of (very large) pixels.

Figure 3.7: Two similar images of Abraham Lincoln.

Figure 3.6 are the same. We'll consider these to be two 3×3 images of (very large) black and white pixels. We'll convert these to points in the 9th dimension. The image on the left is represented by the point $(1, 0, 0, 0, 1, 0, 0, 0, 0)$. Note that the first three coordinates correspond to the top row of the image. The next three coordinates correspond to the middle row of the image, and the last three coordinates correspond to the bottom row of the image. So, the image on the right is represented by the point $(1, 0, 0, 0, 0, 0, 0, 1, 0)$. How far are the images from each other? We again use the Pythagorean theorem, this time to find the distance between points in the 9th dimension. The distance is $\sqrt{2}$. If the images were the same, the distance would be zero.

Now, we are ready to look specifically at the images of Lincoln in Figure 3.4. These images are 632×720 pixels. So, we create points in $(632)(720) = 455{,}040$ dimensions. If we do so, we can easily see that the images in Figures 3.4(a) and 3.4(e) are the same image. Note that we can also use this to find which images are, although not the same, most similar. Let's create our points in 455,040 dimensions for the images in Figures 3.4(b)–3.4(d), 3.4(f), and 3.4(g). When we do this, we find Figures 3.4(b) and 3.4(d) are most similar, although not the same. These images are placed side by side in Figure 3.7.

3.7. A Bit of Pythagorean History

Let's close by discussing the Pythagorean theorem a bit more. The theorem led to a radical idea at that moment in history. Let's consider an isosceles right triangle with sides 1 and 1. Then, the Pythagorean theorem indicates that the hypotenuse has a length $\sqrt{2}$. This is simple enough. The Greeks of the time of Pythagoras stressed that all things in the universe could be reduced into whole numbers or their ratios. Said another way, all numbers could be expressed as fractions. As such, all numbers are rational numbers. So, let's assume that $\sqrt{2}$ is rational. Therefore, we can write $\sqrt{2}$ as a fraction in a reduced form, or

$$\sqrt{2} = m/n, \tag{3.1}$$

where m and n have no common divisors. Squaring both sides of the equation gives us

$$2 = m^2/n^2.$$

We'll then multiply both sides by n^2 and get

$$2n^2 = m^2. \tag{3.2}$$

This tells us that m^2 is even, which implies that m is even. Therefore, m can be written as $m = 2k$ for some integer k. We can rewrite $2n^2 = m^2$ as $2n^2 = 4k^2$. We divide both sides by 2 and find

$$n^2 = 2k^2.$$

So, n^2 is even, implying n must be even. In short, if we assume that $\sqrt{2}$ is rational, then m and n must be even. But remember that we also had $\sqrt{2} = m/n$, where m and n have no common divisors, which means that they can't both be even. This is a contradiction indicating that our initial assumption must be false. Therefore, $\sqrt{2}$ is irrational.

Suddenly, there is a number that cannot be expressed as a fraction. This was mind blowing in those times. So much so, one legend states that the Pythagorean philosopher Hippasus discovered that $\sqrt{2}$ cannot be expressed as a fraction while out at sea. He revealed his radical discovery and was thrown overboard. Math can lead to enlightening results that transform our view of the world — but keep in mind that not all change is accepted easily.

Data exists in many forms. We have the performance of teams and players. We have ratings data. The Pythagorean theorem is a tool to find similar and dissimilar items, and we can do so in 2, 3 or 100 dimensions.

Chapter 4

Fan Excitement

On July 17, 2011, the FIFA Women's World Cup Final was played between USA and Japan. Regulation and extended time ended in a 2-2 draw. In the end, Japan defeated the United States 3-1 on penalty kicks. Twitter exploded at a rate of 7,196 TPS (tweets per second) [6]. How much text is this? We can make a few assumptions to get a rough estimate. The average number of characters per tweet is 33 [49]. So, one can estimate that 237,468 characters per second were zipping through Twitter. Let's compare this to some classic books. The average length of an English word is 4.7. So, approximately 50,000 words per second were being tweeted on that 2011 summer day. This is equivalent to the contents of F. Scott Fitzgerald's *The Great Gatsby* being tweeted every second.

This isn't the only time we see such rates related to sports. Reflecting the worldwide interest in soccer, the 2014 Men's World Cup Final between Germany and Argentina led to a point in the game when a staggering 10,312 tweets per second appeared on Twitter [47]. How do these numbers compare to other Twitter events? For comparison, Beyonce's baby bump revelation led to 8.868 tweets per second in August 2011 [6].

How do numbers get so huge? Such events are truly viral. Let's think about the spread of excitement of a fan base to get context of such a phenomenon.

4.1. Intricate Images

Suppose a fan shares amazing news with another seven fans. Two seconds later, each of those fans has shared the excitement with six new people.

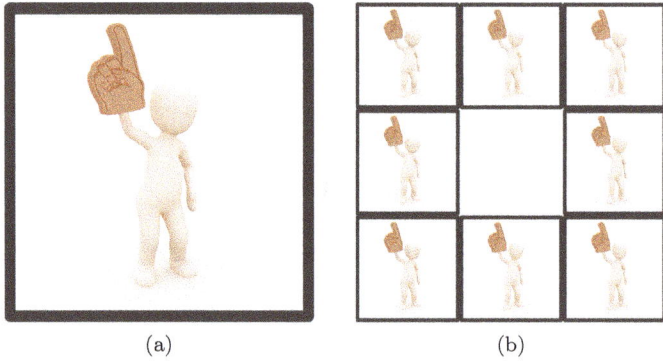

(a)   (b)

Figure 4.1:   We start with one fan in (a). That fan shares the news with seven other fans as seen in (b).

If that pattern continues, how many fans will have the exciting news in 10 seconds?

Let's track the numbers graphically. We begin with you alone as the fan getting the inside scoop on a favorite team, as seen in Figure 4.1(a). To keep track of how many fans have the news, loop through the following steps.

1. Take your current picture and make eight copies of the image reduced in size by a third.

2. Construct a collage by placing the eight images in the configuration seen in the table below:

Image 1	Image 2	Image 3
Image 4		Image 5
Image 6	Image 7	Image 8

3. Does your image look all that different? If not, stop. Else, loop back to step 1 and think of your collage now as your current picture.

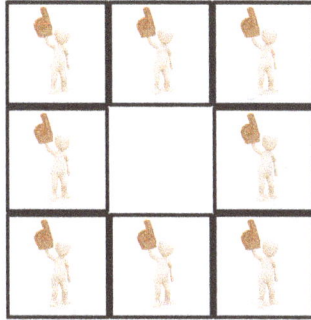

Figure 4.2: One loop of our reduce, cut, and paste algorithm.

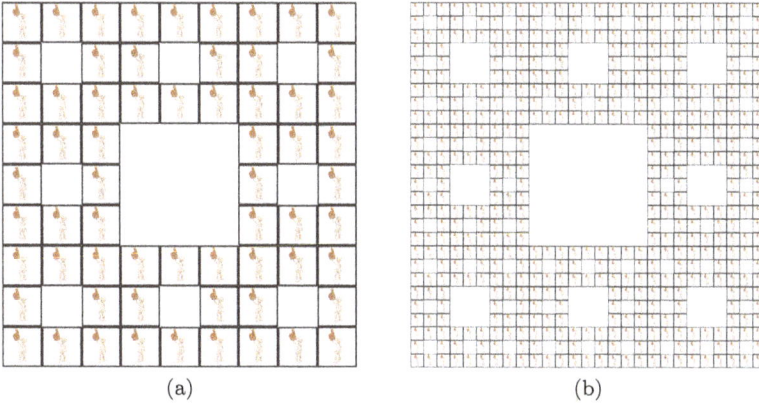

Figure 4.3: More loops of our reduce, cut, and paste algorithm.

The image in Figure 4.2 is what we get after one loop of these steps. This is how many fans know the news after 2 seconds. We haven't reached 10 seconds, so we'll loop again.

If we perform the loop again, we can see how many fans in the know exist after 4 seconds by counting the fans appearing in Figure 4.3(a) and another pass through the loop produces the image seen in Figure 4.3(b).

So, how many people are there? We begin with 1 fan, that's you. After 2 seconds, we have eight fans, one of which is you. After 4 seconds, there are eight sets of eight, giving us 64 fans. After 6 seconds, there are eight sets of 64, or 512 fans. Let's stop and note a pattern.

You begin with $1 = 8^0$ fans. Then, you have $8 = 8^1$ fans. After this, there are $64 = 8^2$ fans. Can you see the pattern? After 6 seconds, there will be $512 = 8^3$ fans. And 10 seconds leads to $8^5 = 32{,}768$ fans. This is very fast growth. To see how fast, how many fans appear after 24 seconds? This would be $8^{12} = 68{,}719{,}476{,}736$. If we filled each square with a face from someone on the planet, then we'd have more squares than faces! That's fast growth — in fact, exponential growth!

Maybe the key to this fast growth is each person sharing the news with seven people. Let's explore the impact of each person sharing the news with two people. This time, we follow the following steps:

1. Take your current picture and make three copies of the image reduced in size by 50%.
2. Construct a collage by placing the three images in the configuration seen in the table below:

Image 1	
Image 2	Image 3

3. If your image looks essentially the same, stop. Else, loop back up to step 1 and think of your collage now as your current picture.

The image in Figure 4.4 is what we get after one loop of these steps. This looks significantly different from the image of the single fan. So, we loop again.

If we perform the loop again, we get the image on the left in Figure 4.5 and another pass through the loop produces the image on the right.

It's a matter of taste as to when to stop. In Figure 4.6, we see the image after six loops versus seven loops. Very little changes, which would cause us to stop! How fast are things growing, however? The first loop produces three fans. The second loop creates $9 = 3^2$. The third loop has $3^3 = 27$. If you looped 10 times, the image would have 59,049 fans. If you looped 21

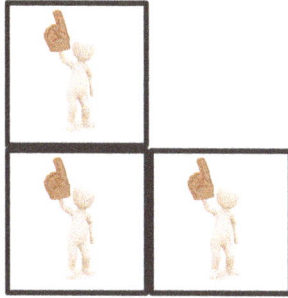

Figure 4.4: One loop where each fan informs two other fans.

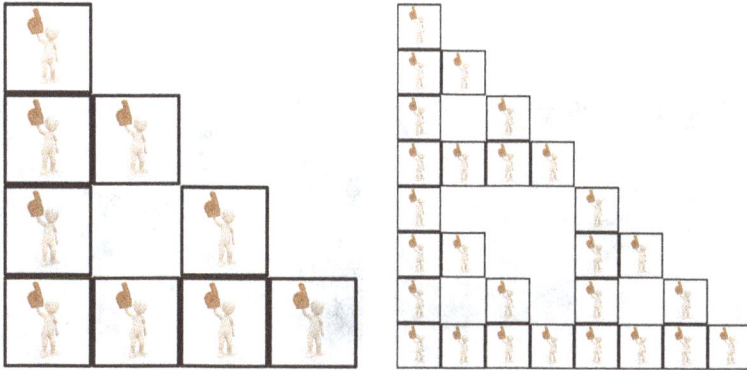

Figure 4.5: Additional loops of a fan informing two new fans about exciting news.

times, then you'd have 10,460,353,203 fans, which is again larger than the population of the earth.

Sharing our news can be captured with the function $f(x) = b^x$, where $b = 8$ when each fan shares the news with 7 fans. Using this formula, we can directly compute the number of fans in Figure 4.1(a) as $f(0) = 8^0 = 1$, and that in Figure 4.1(b) as $f(1) = 8^1 = 8$. Similarly, the number of fans in Figures 4.3(a) and 4.3(b) equals $f(2) = 8^2 = 64$ and $f(3) = 8^3 = 512$, respectively. Such a function is said to be exponential and have exponential growth. As we've seen, this is very fast growth and it does help us to capture behavior in our world.

Before moving on, let's look at an important feature of the square and triangle we've created. Note that the images contain copies of the larger

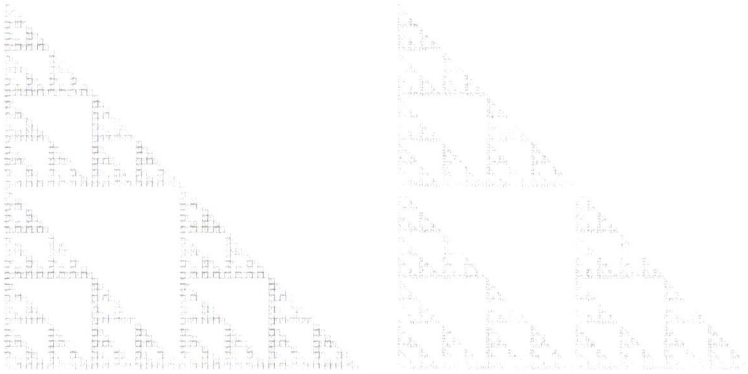

Figure 4.6: What if we looped six or seven times where a fan produces two more fans? Here are the resulting images.

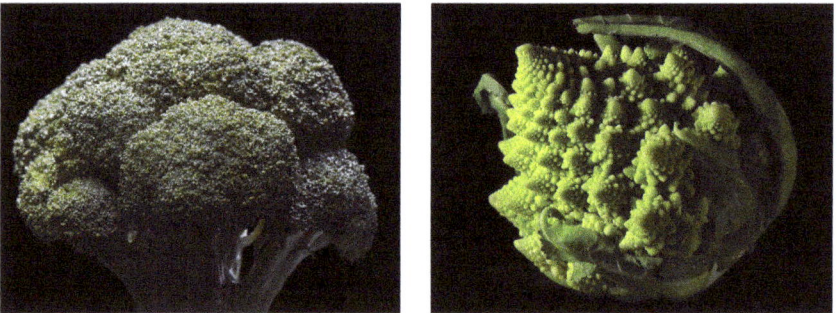

Figure 4.7: Broccoli, in its various forms, supplies a real-life object with fractal-like attributes.

image. Zooming into an object and seeing similarities to the whole is an important property of what are called fractals in mathematics. An object has self-similarity if it looks the same as or similar to itself under increasing magnification.

Broccoli, as seen on the left in Figure 4.7, exhibits the properties of self-similarity. Break off a stalk and it looks like the larger stalk of broccoli — just at a different scale. Another example is Romanesco broccoli as seen on the right in Figure 4.7.

How does this level of growth, which is simply multiplying by the same number over and over and over again, connect to the world around us? If someone claims to have broken the Internet, they could also claim to have, quite likely, witnessed exponential growth. Things with small starts can have huge, record-breaking ends.

4.2. Math of the Instagram Egg

Let's model something going viral on social media to get a sense of how exponential growth can fit. We'll take the Instagram egg which was posted on the @world_record_egg account on the social media platform Instagram. It took only days for the egg to break records. The photo was originally taken by Serghei Platanov, and was simply a brown egg sitting upright against a white background as seen in Figure 4.8. Simple enough! On January 4, 2019, the @world_record_egg account was created, and there was posted an image of the egg. The caption read, "Let's set a world record together and get the most liked post on Instagram. Beating the current world record held by Kylie Jenner (18 million)! We got this." The image reached 18.4 million likes in 10 days to become the most-liked Instagram post of all time. It only took an additional 48 hours for the post to surpass 45 million likes to become the most-liked online post (on any media platform) in history.

Let's use the ideas of this chapter to model this process. We'll begin with one person liking the post about the simple egg. After one day, seven new people like the post, giving us a total of eight likes. The next day, there would be 49 new likes, giving the image a total of $49 + 7 + 1 = 53$ likes. Not

Figure 4.8: This image of an egg would break world records.

much of a sensation — yet! After nine days, there would be over 47 million likes, if that rate were maintained.

The egg received 18 million likes in 10 days, so this model grows too quickly. We need only go to 6 new people liking the post at each stage rather than 7 to reflect the 10-day growth rate of the Instagram egg. To see this, we again begin with one person. After one day we have $1 + 6 = 7$ total likes, and after two days, we have $1 + 6 + 36 = 43$ total likes. On the ninth day, we have just over 12 million likes. Therefore, before the 10th day ends, we'd surpass 18 million likes. Even so, this growth rate is too fast to hit 45 million after an additional 48 hours. With this model, we'd surpass 18 million and 45 million likes on the 10th day. In fact, we would end the 10th day with over 72 million likes. One way to adjust the model would be to still have quick growth but not exponential. We explore this in Section 4.3.

4.3. Not So Fast

How would we know if we have exponential growth? We could also have exponential decay, as we can also have very quick declines, too. In the previous section, we saw how exponential growth could emulate a phenomenon. We can also see how well exponential growth fits existing data. For this, let's look at a model of exponential decay in money won versus average strokes on the Ladies Professional Golf Association (LPGA). You can see the 2018 data for the top 10 ranked players in Table 4.1.

Table 4.1:   2018 data for the top 10 ranked LPGA players.

Rank	Name	Official money	Average strokes
1	Ariya Jutanugarn	$2,743,949	69.415
2	Minjee Lee	$1,551,032	69.747
3	Sung Hyun Park	$1,498,077	70.646
4	Brooke M. Henderson	$1,473,247	69.99
5	Nasa Hataoka	$1,454,261	70.103
6	So Yeon Ryu	$1,438,850	70.102
7	Sei Young Kim	$1,369,418	70.265
8	Carlota Ciganda	$1,244,610	70.088
9	Lexi Thompson	$1,223,748	70.014
10	Jin Young Ko	$1,159,005	69.806

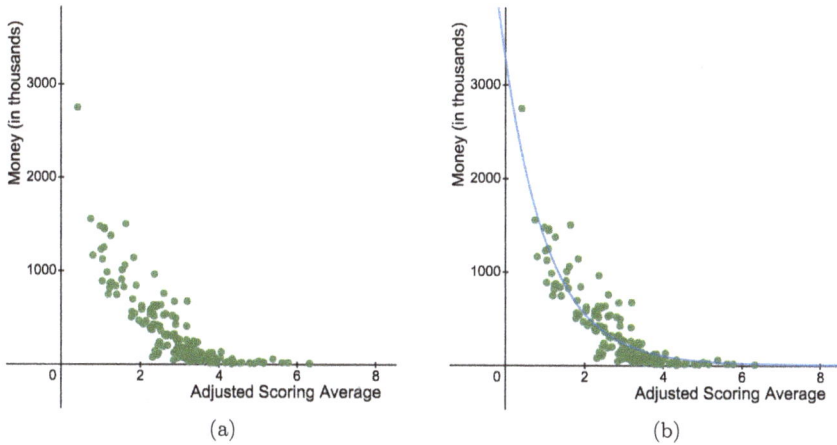

Figure 4.9: 2018 data for about 160 top-ranked LPGA players (a), with an exponential best fit line in (b).

We clearly see a significant drop-off from Ariya Jutanugarn's $2,743,949 to Minjee Lee's $1,551,032. However, it can be hard to discern the rate of decline from that point. So, let's take data from about 160 players. We see the data in Figure 4.9(a). Note that for the x-axis, we have subtracted 69 from the player's average strokes. From here, we find the curve of the best fit and find that $y = 3293e^{-0.8923x}$ is the best fitting curve. So, we see the exponential decay of money as it relates to the average number of strokes of a player on the tour.

Looking at the curve, the points appear to hover about the curve. Could the points hover more snuggly around a quadratic than an exponential curve? In this case, no. When software gives a curve of best fit, it usually also gives you an R^2 value for the curve. If $R^2 = 1$, then we have a perfect correlation between the data and the curve. The closer R^2 is to 1, the better the fit. We find that for the LPGA data, an exponential fit is better than a quadratic polynomial.

If you use a search engine to query on "exponential growth" in news outlets, you'll likely find articles where growth is clearly fast but questionably exponential. Still, exponential growth can be a good model for short periods of time. What about data that isn't exponential? For this, let's look at the team value of the Golden State Warriors NBA team as determined by Forbes. Let's take $x = 0$ to be the value of the team in 2010 (so $x = 9$

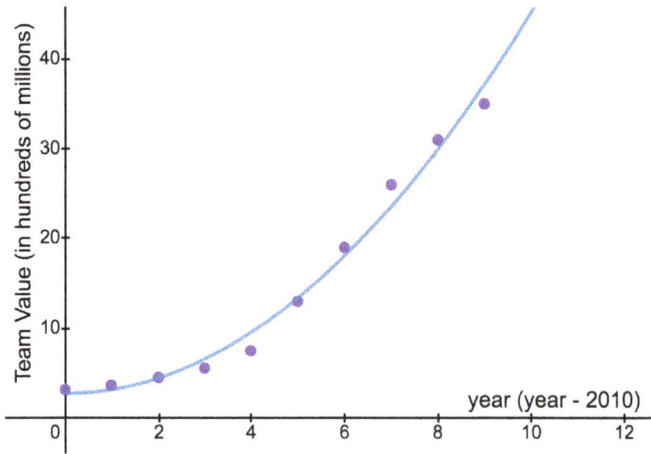

Figure 4.10:   Team value of the Golden State Warriors between 2010 ($x = 0$) and 2019 ($x = 9$).

is the value in 2019). Then, the team value in 2010 was \$315 by Forbes and 3.5 billion in 2019. The graph for every year between 2010 and 2019 is shown in Figure 4.10. Clearly, the team had impressive growth in value. Is an exponential the best fit? In this case, it isn't. If we fit both an exponential and a quadratic, we find that in this case, a parabola is a better fit, as depicted by the blue curve in Figure 4.10. So the Warriors had significant growth in team value over this period. Even so, it is not best modeled by exponential growth.

Exponential growth is fast — really fast. The next time you hear about a social media phenomenon, you may be witnessing the effects of exponential growth.

Chapter 5

Predicting the Future

Paul the Octopus, who lived for just over two and a half years, was known worldwide to be an animal oracle. To divine the result of an upcoming soccer match, Paul would be presented two identical boxes containing food, except that they were decorated with different team flags of the competitors of an upcoming soccer match. Paul's predicted winner was discerned from the box from which Paul ate first.

Paul lived in the Sea Life Centre in Oberhausen, Germany. Paul mainly predicted the outcomes of the international matches involving the German national team. Paul correctly predicted the winner in four out of Germany's six Euro 2008 matches, and in all seven of their matches in the 2010 World Cup. He also correctly predicted Spain's win in the 2010 FIFA World Cup final as shown in Figure 5.1. At his death in October 2010, Paul had predicted the outcome of 12 of 14 matches, giving him an 85.7% success rate [28].

In this chapter, we'll discuss some techniques, instead of hoping for an animal oracle to help us gain insight into teams and their possible future performance.

5.1. Pythagoras' Crystal Ball

The greats of sports transform the game. Bill James is undoubtedly one of the greats of baseball. Yet his greatest asset wasn't a great arm or bat, but mathematics. He is now, as the staff of the Red Sox, using mathematics to enhance the team.

Keep in mind that even the giants of sport were younger, less-experienced versions of themselves. Michael Jordan still had to learn to shoot, pass and

Figure 5.1: Paul the Octopus correctly predicts the win by Spain over Germany in the 2010 FIFA World Cup semi-final.

dribble. Keep in mind that Jordan was always a future NBA star, but that's something we know only now. As a sophomore in high school, Jordan didn't make the basketball team in Wilmington, North Carolina. He did his junior year and had several 40-point games.

In a similar way, it is unlikely that Bill James knew the eventual impact of his work when he was first creating the formulas that served as a foundation of what is now called sabermetrics. James would later be named one of the most influential people in the world by *Time* magazine in 2006. Many of his baseball articles were penned while he worked nightshifts as a security guard at the Stokely-Van Camp's pork and beans cannery. Let's see how to develop Bill James' Pythagorean expectation.

First, let's learn about the formula. A fraction will estimate the percentage of games a team will win. Specifically,

$$\text{winning percentage} = \frac{(\text{runs scored})^2}{(\text{runs scored})^2 + (\text{runs allowed})^2}.$$

Let's use this formula on the 2002 Oakland A's. The team scored 800 runs and allowed 654 runs during the regular season. So, the Pythagorean expectation equals $(800)^2/(800^2 + 654^2) = 0.5994$, indicating that the team is expected to win 59.94% of their games.

The 2002 A's team played 162 games and 59.94% of 162 is 97.1. The team won 103 games and thus exceeded their expectation by 5.9 games.

Teams that exceed their Pythagorean expectation are sometimes called lucky.

Note that you can also take a team's stats for runs scored and allowed during a season and project how many games they will win by the end of the season. In early June 2017, the Yankees had won 44 games and lost 39. They'd scored 464 runs and given up 362. Using the Pythagorean expectation, they were expected to win 51.6 games but had only won 44. They had been unlucky. Did their luck change? One way to measure it is to project out to 100 games. Our formula predicts that the Yankees should have won 62.2% of their games in early July. At the 100 game milestone, the Yankees had won 54, still struggling on the unlucky side of the formula.

Can we develop such a formula for other sports? Let's try basketball and with it change some terminology. Points rather than runs are scored. Let's find an exponent in the Pythagorean expectation that can work in basketball. We currently don't know its value, so it will be treated as our unknown. Our equation is

$$\text{winning percentage} = \frac{(\text{points scored})^x}{(\text{points scored})^x + (\text{points allowed})^x}.$$

How do we determine a suitable exponent? One approach is to find the exact exponent for each team. Let's do this for the 2012–2013 NBA season. For example, consider the Miami Heat. They scored 8436 points, allowed 7791 points, and won 66 of their 82 games. So, the Pythagorean expectation for them is $8436^x/(8436^x + 7791^x) = 66/82 = 0.8048$.

There are two ways to find x. First, we can apply algebra techniques and solve. Warning: this can become pretty involved! Just look at the exact answer:

$$x = \frac{-3\log(2) + \log(3) + \log(11)}{2\log(2) - 2\log(7) + \log(19) + \log(37) - \log(53)}.$$

Alternatively, one can make a guess. Here is how this works. Let's lean on baseball and take $x = 2$. Now: $8436^2/(8436^2 + 7791^2) = 0.5397$. This estimate is too low. We'll also want an exponent that produces an estimated winning percentage that's too high. Using $x = 20$, we find that $8436^{20}/(8436^{20} + 7791^{20}) = 0.8307$, which is indeed too high. So, the value for x lies somewhere between 2 and 20. We can use a method known as the bisection method and simply choose the value halfway between 2 and 20, which equals 11, giving us $8436^{11}/(8436^{11} + 7791^{11}) = 0.7057$, which is too low. So, we now know that the value for x is between 11 and 20. We again pick the value in the middle and test it. In this way, we keep finding two values for x: one that produces an estimate that's too high and one too low. When we get a value that produces a value for the corresponding Pythagorean expectation that's close to 0.8048, we have found our desired value for x, which for the Miami Heat equals 17.81.

A value of 5.87 works for the 2012–2013 Houston Rockets and 33.35 for the Golden State Warriors. We need to know the value for x for every team. With so much disagreement, how can we find anything? A few approaches could be taken. We can simply take an average of the computed x values for all the teams. If we do this, we find that x is about 12. We could also find the median, which equals 15.

Generally, the exponent in the Pythagorean expectation for basketball is taken to equal about 14. Our average was screwed a bit low due to outliers like the Utah Jazz. They scored 8038 points and allowed 8045. The Pythagorean expectation estimates that they should have lost more games than they won. Think about it. They scored less points than they allowed. But they won 43 and lost 41.

If we repeat this process for baseball, will we find an exponent of 2 like Bill James? We'll run the numbers on the 2013 regular season of Major League Baseball. If we take the average of the computed exponents for all the teams, we find that $x = 2$! Try it yourself on 2013 or another season. Use pencil, paper, and a calculator and be like Bill James and start crunching the numbers!

5.2. Lies and Correlations

Sometimes, we can think that we have made a prediction when we haven't. In sports analytics, a key to successful work is to, frankly, doubt yourself. Confirm your results, else your work can support the saying "Lies, damned lies, and statistics," which was a phrase popularized in the United States by Mark Twain in "Chapters from My Autobiography" published in the *North American Review* in 1906. A place where it can be easy to misrepresent a result is to interpret correlations as causations. In sports, this can lead to superstitions.

In college, Michael Jordan led the University of North Carolina Tarheels to the NCAA Championships in 1982. Wanting to repeat such success, Jordan kept wearing the shorts. The Chicago Bulls won six NBA championships with Michael Jordan. Through it all and in every game, the five-time MVP wore his University of North Carolina shorts under his uniform.

Jordan isn't alone in having rituals. Kevin Rhomberg played just 41 Major League Baseball games for the Cleveland Indians from 1982–1984. That short stint of games is quite memorable, mainly for Rhomberg's superstitions. Rhomberg was a good hitter with a 0.383 batting average. Rhomberg differed from other players due to a self-imposed rule not to make right turns while on the field since base running only involves turning left. For Rhomberg, the rule wasn't just for base running. If a ball was hit to the left field in a way where any other player would turn right, Rhomberg would make a full circle, turning left, of course.

Such superstitions can be funny but also underscore a tendency, we each have to find patterns. For example, we might see or compute a correlation. What can be humorous is our own tendency to imagine why one factor would cause another. Some correlations are directly connected, with one causing the other. Others are not.

A fun exercise is to purposely find correlations that are difficult to imagine as casual. Even so, you might find yourself creating a reason. Here are some correlations:

- Did you know that we might have found an answer to population growth in the United States? There is a positive correlation between the popularity of the name "Claire" and the total population of the United States between 1975 and 2015.
- Did you know Clemson fans might want to root for MLB sluggers? There is a positive correlation between the number of wins by the Clemson University football team per season and home runs hit in the MLB per year between 2005 and 2019.
- Talking politics? MLB pitching may be at the center of how to control our national debt. There is a positive correlation between the amount of

the US government's total debt since 1950 and the amount of home runs hit in each MLB season.

- Watching the NFL? Is it taking too long with penalties? Maybe too many people are thinking about buying a bed on Black Friday. There is a positive correlation between the general interest in beds and the number of penalty flags thrown in an NFL football game between 2009 and 2018.

These were created by my students during a data science class at Davidson College. One student even found that having less people taking the train in Germany could lower the cost of college. There is a strong positive correlation between the cost of attendance at Davidson College and the total number of train passengers in Germany from 2009 to 2018.

How did my students construct such correlations? Let me explain how I made one. I learned that a paper in the *New England Journal of Medicine* in 2012 has found a correlation between chocolate consumption in a country and the number of Nobel Laureates produced by that country [38]. Given this, I wanted to see if I could create a correlation between eating chocolate and studying mathematics.

What connection might we make? First, I found the number of doctorates in engineering and science awarded in the United States via the National Science Foundation. Then, WolframAlpha reported the chocolate consumption (measured in trillions of dollars spent on buying the sweet) in the United States. Both results are shown in Table 5.1.

Table 5.1: The number of doctorates in engineering and science awarded in the United States and chocolate consumption (measured in trillions of dollars) in the United States.

Year	No. of doctorates	Chocolate consumption
2002	24608	9.037
2003	25282	9.52
2004	26274	10.13
2005	27986	10.77
2006	29863	11.39
2007	31800	11.95
2008	32827	12.37

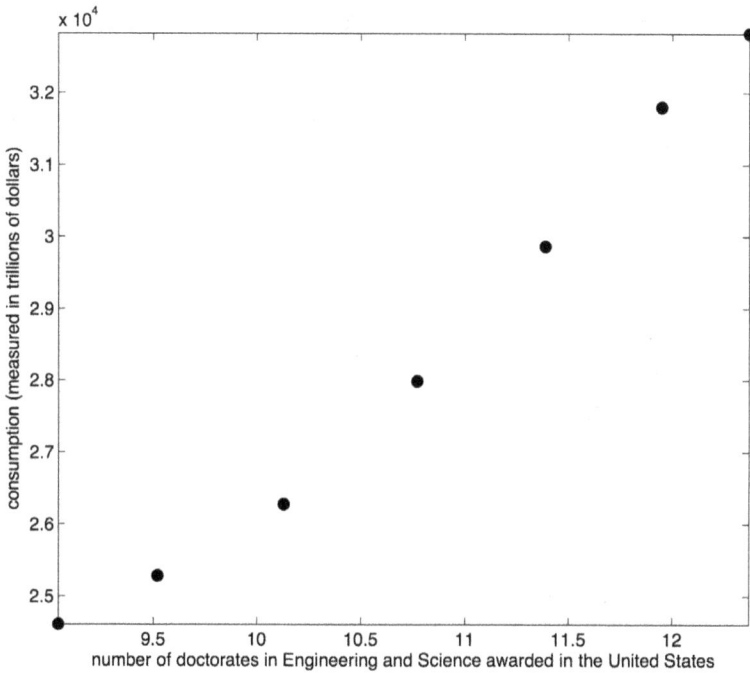

Figure 5.2: Plot of the chocolate consumption (measured in trillions of dollars) in the United States versus the number of doctorates in engineering and science awarded in the United States.

The first step is to plot the data. Let the x-axis be the dollars spent on chocolate and the y-axis be the number of doctorates awarded in science and engineering in the United States. From the graph in Figure 5.2, we can see that the data is close to linear, so we expect a correlation, which can be found using Excel or Google Sheets. In the end, for this data, the correlation is 0.99, indicating a strong correlation between the number of doctorates in science and engineering and chocolate consumption.

Does correlation mean causation? Should we put a chocolate in the hands of every STEM student to increase our national standing in this area? If a box of chocolates is in front of me, I would like to think that eating chocolate will improve my mathematics. Yet, correlation doesn't imply causation, so I'm also not surprised when I seem to perform about the same even after the box of chocolates has become half full.

5.3. Regress to a New World Record

A major tool in predictive analysis is using regression or least squares. Regression fits data to a line. To understand what this means, let's think about what regression would give us for the three blue points in Figure 5.3. If we have a line, we are interested in the least-square error. Note that E_1 equals the distance from (x_1, y_1) to the point on the line with the same x value. The least-square error equals the sum of the squares of the vertical distances from the data points to their corresponding point on the line. For Figure 5.3, the least-square error for this line would equal $E_1^2 + E_2^2 + E_3^2$. Regression would help find the line with the smallest least-square error.

So, when is linear behavior a good approximation to input data? Much like the previous section, if we can graph the data, that's a great place to start. Note how the data in Figure 5.4(a) can be seen as a cloud of points hovering around some underlying line, indicating that linear regression may work well. By comparison, the data in Figure 5.4(b) does not share this property. When you want to create a regression line, you can plug the data into many

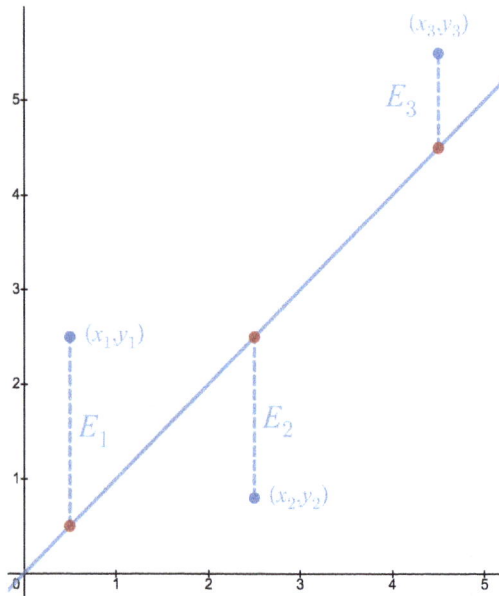

Figure 5.3: Fitting data to a line is a useful tool in data analytics.

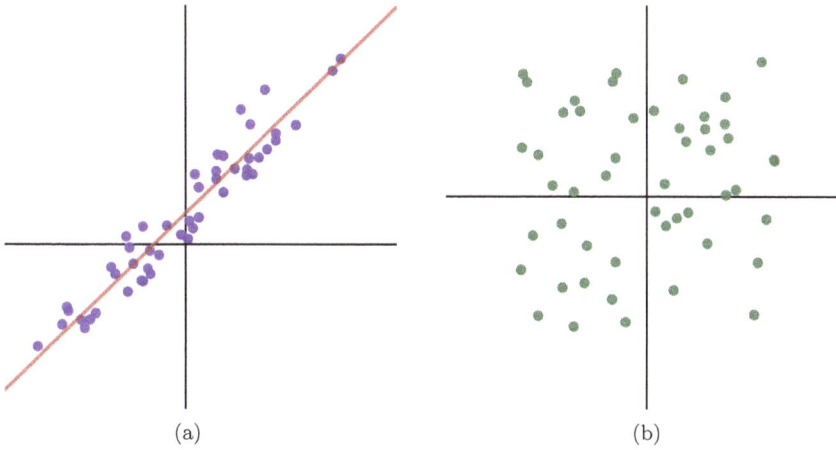

(a)   (b)

Figure 5.4:   Some data has a linear trend (a) and some does not (b).

calculators or spreadsheets and quickly find the line of best fit. While it's easy to find, good data analysis starts by ensuring that the tool fits the job.

So, let's use regression to analyze world record times in the 25 meter freestyle sprint in women's swimming. First, let's see the data.

Time	Name	Nationality	Date	Years
53.46	Franziska van Almsick	Germany	January 6, 1993	0
53.33	Franziska van Almsick	Germany	January 10, 1993	0.011
53.01	Jingyi Le	China	December 2, 1993	0.904
52.8	Therese Alshammar	Sweden	December 10, 1999	6.929
52.17	Therese Alshammar	Sweden	March 17, 2000	7.197
51.91	Libby Lenton	Australia	August 8, 2005	12.595
51.7	Libby Lenton	Australia	August 9, 2005	12.597
51.01	Libby Trickett	Australia	August 10, 2009	16.603
50.91	Cate Campbell	Australia	November 28, 2015	22.907
50.77	Sarah Sjöström	Sweden	August 3, 2017	24.589
50.58	Sarah Sjöström	Sweden	August 11, 2017	24.611
50.25	Cate Campbell	Australia	October 26, 2017	24.819

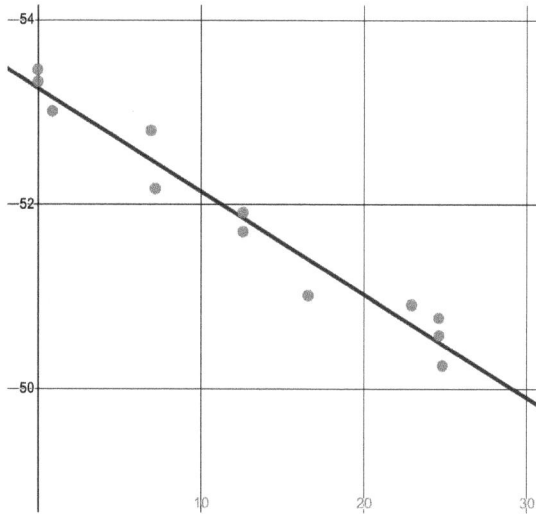

Figure 5.5: World record times in the 25 meter freestyle sprint in women's swimming and a least squares fit to the data.

We need to decide how we'll graph the data, which is produced in the last column of the table. In particular, we are going to let the x-axis represent the date that a world record time (presented by the y-axis) was set. We'll let January 6, 1993 have an x value of 0. Then all other dates are the number of years that passed from the first world record time by Franziska van Almsick (which was set to 0). Regression produces the line of best fit, $y = -0.112x + 53.2602$, as seen in the graph of the line and data point in Figure 5.5.

This equation tells us that the world record winning time tends to decrease by 0.112 seconds per year. Note that this is a least-squares line, so $y(0) = 52.26$, even though Franziska van Almsick set the record on January 6, 1993 at 53.46. We can read several things from our line of best fit. Points below the line represent bigger jumps in the world record time than predicted by the regression line. For example, the record set by Australia's Cate Campbell is below the line. Interestingly, Therese Alshammar's first record of 52.8 is above the line, while her next record of 52.17 is below. Both represent historic moments as she set a world record, yet her performance in the year 2000 signified a much more significant drop in the world record.

We can use our least-squares line in other ways. When will the world record be 49 seconds, at least according to our regression? The y-axis represents world record times. So, we are interested in asking when $49 = -0.112x + 52.26$, or $x = 38.04$, which corresponds to early 2031. Now, first, this is an approximation. As we saw, it wasn't correct for year 0, which is the January 6, 1993 record of 53.46. Further, and more importantly, there is the question of whether the data will continue this linear trend. Keep in mind that we could also ask when someone would swim the 25 meters in 2 seconds, which isn't possible. At some point, it is likely that times will begin to plateau as some limit point of human performance approaches.

5.4. Google's Billion-Dollar Ranking

If you search the web on the term "sports analytics", you get a list of web pages. We tend to start at the top of the list and work our way down. This is partly why Google has become Google. We no longer need to look at the 20th or 50th page of search engine results. Google has, as part of its algorithm, a measurement of the quality of a web page. The company has been so successful that Google is today not just a noun. "Google it" is one response many of us have to information we may not know.

How is it done? Google modeled the movement of a random surfer moving through the Internet. To do so, the surfer needs rules. On any web page, if there are no links to click, then the surfer can go anywhere on the web. Yes, anywhere, with equal probability. If there are links on the page, then the surfer follows these rules:

- 85% of the time, the surfer clicks one of the links on the web page. The surfer is equally likely to pick any link.
- 15% of the time, the surfer goes anywhere on the Internet.

That's it. This simple model is an important part of what catapulted Google to become the search engine giant.

The quality of a page is measured as the percentage of time that a random surfer would spend on that Web page if the surfing occurred forever. How does Google ever determine the quality of any page if the surfer is going to be doing this from now to beyond the end of time? Google has guaranteed that following these rules, the probabilities will settle down into what is often called a steady state.

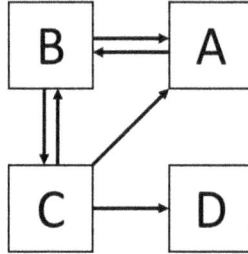

Figure 5.6: Web network.

Suppose we find the steady-state probabilities for the network in Figure 5.6. Note that web page A has a link to web page B. Web page B has links to A and C. Web page C has links to all the other web pages, and D doesn't have any links, which is often called a dangling node. If the random surfer moves along this web network following Google's rules, the surfer would visit web pages A, B, C, and D about 28.4%, 36.8%, 22.1%, and 12.7% of the time, respectively. Therefore, web page B has the highest ranking in terms of the quality of the pages. After web page B comes A, C, and then finally D.

Part of the brilliance of this model is that it captures enough realism that the results of Google searches make sense to us. Yet, the model has theoretical foundations that guarantee rankings regardless of the structure of the World Wide Web. In the next chapter, we'll look at ranking in the context of March Madness and show that ranking can help us to gain insight into upcoming sporting events. Google's ranking, in a sense, works to predict how we'd measure the quality of web pages.

Chapter 6

Your Number One

Every March, a large amount of national attention in the United States turns to NCAA Division I men's basketball and the tournament known as March Madness. This single elimination tournament begins with 68 teams from approximately 350 schools and crowns a national champion with the only team that ends the tournament without a loss. The initial pairings of teams are announced on Selection Sunday and visualized as part of a tournament bracket, a portion of which is shown in Figure 6.1. The match-ups shown on the solid lines on the left side of the bracket are the announced games and the Madness of March often comes in predicting the winners of each game (dashed lines) to see who advances to eventually win the tournament. By the following Thursday, bracket predictions must be complete and entered in pools in which friends, colleagues, and even strangers compete for pride, office pool winnings, or even thousands of dollars. In fact, in 2014, Warren Buffett ensured a billion dollar prize for anyone who could complete a perfect bracket for the tournament. By the end of the second round (or 48 games) none of the over 8 million brackets had perfectly predicted the 2014 tournament. As the tournament progresses, game results either uphold or overturn the predictions of millions of fans. For instance, for the match-ups shown in Figure 6.1, Connecticut beat St. Joseph's, which matches the prediction, and Villanova beat Milwaukee, which goes against the prediction. The actual results from 2014 are shown in Figure 6.2.

Let's see how to use mathematics to pick a bracket, which we'll consider for March Madness but can be adapted to other sporting events like the FIFA World Cup. It is important to distinguish between ratings and rankings. A *rating* is a value assigned to each team from which we can then form a

63

Connecticut

St. Joseph's ──── Connecticut

Villanova ──── Connecticut

Milwaukee ──── Milwaukee

Figure 6.1:   A small portion of a 2014 March Madness bracket.

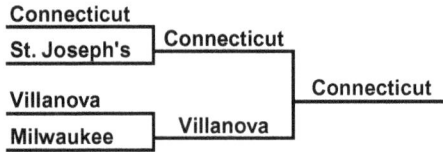

Connecticut

St. Joseph's ──── Connecticut

Villanova ──── Connecticut

Milwaukee ──── Villanova

Figure 6.2:   A small portion of the 2014 March Madness tournament bracket results.

ranking, an ordered list of teams based on a chosen rating method. So, we'll define methods that create ratings for all the teams and then sort the ratings in descending order to create a ranked list from the first to the last place. For a math-based bracket, a team with a higher rating is predicted to win each game.

6.1. Winning Percentage

A natural ranking method is to use winning percentage. As such, a team's rating is computed as each team's ratio of the number of games won to the total number of games played. However, a simple system like this includes only information about each game's end result and can lead to misleading information. For instance, if Team A plays only one basketball game and they win, then their winning percentage rating would be 1. If Team B plays 20 basketball games and goes undefeated, their winning percentage rating would also be 1. Intuitively, Team B is the more impressive team since they went undefeated for 20 games, while Team A won only once. But they both have the same winning percentage, so this rating system would produce a tie when ranking these two teams. Yet, problems also arise even if Teams C and D both played 20 games, but C remained undefeated by playing a difficult schedule composed of the most talented teams in the league, while D played only the weakest teams. While D may be a talented team, that fact is not readily apparent from their schedule. A fundamental difference in this schedule is the strength of C's and D's opponents.

Let's look at the 2014 March Madness tournament, given that it had a billion dollar prize associated with it. If we use winning percentage, Wichita State would be predicted to win the tournament, with a rating of 1 from their undefeated season. Even the selection committee didn't take that record to indicate overall dominance as Wichita was ranked third in the tournament. In the end, they would win only one game. The team with the next best winning percentage was Stephen F. Austin. They were ranked 50th in the tournament. They did have an initial upset when they beat VCU but then lost in the next round. In the end, winning percentage would correctly predict 50.8% of the games in the 2014 tournament.

6.2. All About Wins and Losses

To integrate strength of schedule or the quality of one's opponent into a rating system, we'll use linear systems. An example of a linear system is $x + y = 6$ and $x - y = 4$. What values for x and y make both equations true? If you add the equations together, you get $2x = 10$ or $x = 5$. Since $x + y = 6$, $y = 1$. For bookkeeping, we'll use matrices. This system of two unknowns and two equations becomes

$$\begin{pmatrix} 1 & 1 \\ 1 & -1 \end{pmatrix} \begin{pmatrix} x \\ y \end{pmatrix} = \begin{pmatrix} 6 \\ 4 \end{pmatrix}.$$

If matrices are new to you, mainly note that any matrix system can be made into a linear system. The square matrix, where the number of rows equals the number of columns, on the left holds the coefficients of our linear system. The vector, with one column, on the right holds the values on the right-hand side of our system. So, the matrix system

$$\begin{pmatrix} 1 & 2 & -1 \\ 0 & 1 & 2 \\ 3 & 1 & 2 \end{pmatrix} \begin{pmatrix} x \\ y \\ z \end{pmatrix} = \begin{pmatrix} 3 \\ -4 \\ -1 \end{pmatrix}$$

becomes the linear equations $x + 2y - z = 3$, $y + 2z = -4$, and $3x + y + 2z = -1$. If you add equations and eliminate variables, you'll find $x = 1$, $y = 0$

and $z = -2$. Note that you can also go to WolframAlpha.com and type

$$\text{solve } x + 2y - z = 3, \quad y + 2z = -4 \quad \text{and} \quad 3x + y + 2z = -1.$$

A course or on-line resources in linear algebra can give you more information on solving matrix systems. Ours will be much larger and will be solved by computers as it is highly impractical to do them by hand.

Our first linear system will be created using the Colley method, as introduced in [27]. Methods using linear algebra differ from winning percentage. Linear algebra integrates interdependence of the ratings. Specifically, it is by computing all of the ratings at one time that we can tackle the issue of strength of schedule.

The Colley method attains this through its derivation. The method starts with the assumption that every team's rating is 0.5 at the beginning of a season. It's interesting to note that even this assumption gives superior results to winning percentage. Before any games are played, the Colley method rates all teams with the value of 0.5. No games have been played, so all teams are considered equal. Winning percentage, however, does not yield a number, but only the ever-troublesome 0/0.

As games are played, the Colley method adjusts its rating to incorporate game results. A team's rating is also affected by the rating of every team it plays. We would now distinguish between Team C with its undefeated season against very strong opponents and Team D with its undefeated record against very weak opponents. Colley would rate Team C higher than Team D.

An important part of the Colley method is its omission of point differentials. So, a rout by 25 points counts the same as a win in double overtime by 1 point. Some may argue that all wins are not created equal. However, it can also be that scores create random noise in a system as some games are close until the final minute or a rout is reduced with the second string playing a large portion of the game. Soon, we'll see a method, called the Massey method, that integrates point differentials.

In the Colley and Massey methods, the ratings of a team's opponents contribute to the team's rating. Mathematically, this results in a system of n equations and n unknowns when computing the ratings of n teams. For March Madness, we take the results of all NCAA Division I teams, which results in roughly 350 unknowns.

Let's see how to create the matrix system for the Colley method with a fictional series of games between teams that play NCAA Division I men's basketball. To keep our system reasonably small, we'll restrict our discussion to four teams: College of Charleston, Furman University, Davidson College, and Appalachian State University. We'll represent the records of the teams by a graph where an arrow represents a game and will point from the winning team to the losing team. Although not needed for the Colley method, the value written next to the arrow indicates the difference between the scores of the winning and losing teams. In Figure 6.3, we see, starting from the upper left and moving clockwise, the College of Charleston, Furman, Davidson, and Appalachian State.

We begin with the unknowns. We need a rating for every team. Then, we assign each row to a team. Any order will do. Let's assign schools to rows of the matrix starting from the upper left and moving clockwise in Figure 6.3. So our unknowns will be C, F, D, and A for the College of Charleston, Furman, Davidson, and Appalachian State, respectively. This helps us form the square matrix in the linear system. The first row corresponds to the College of Charleston, the second to Furman, the third to Davidson, and the last is Appalachian State. The columns have the same ordering.

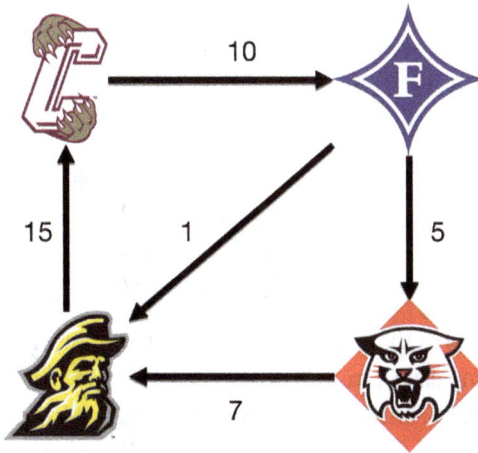

Figure 6.3: A fictional season played between NCAA basketball teams. An arrow points from the winning team to the losing team. The number next to the arrow indicates the difference between the winning and losing scores.

Our square matrix will have four rows and four columns, since we have four teams. Now, we need to know how to fill in the diagonal elements, which are depicted in light red below. The off-diagonal elements of the matrix are depicted in light blue.

$$
\begin{pmatrix}
\blacksquare & & & \\
& \blacksquare & & \\
& & \blacksquare & \\
& & & \blacksquare
\end{pmatrix}
$$

Each diagonal element equals the number of games that the team associated with that row has played. The first row is the College of Charleston. From Figure 6.3, we see that they played two games; remember that each arrow (in or out) of a team logo represents a game. So, the element in the first row and first column of the matrix equals 4. Furman played three games, so its diagonal element becomes 5. The off-diagonal elements are also simply a matter of looking at the arrows. Each off-diagonal element corresponds to the number of games played between the teams associated with the row and column of that element. Let's look at the element in the first row and second column of our square matrix. This corresponds to the College of Charleston (given that we are in the first row) and Furman (given that we are in the second column). The off-diagonal element equals the number of games the associated teams played against each other multiplied by -1. So, the element in the first row and second column of our square matrix equals -1. Doing so for every team yields the following matrix:

$$
\begin{pmatrix}
4 & -1 & 0 & -1 \\
-1 & 5 & -1 & -1 \\
0 & -1 & 4 & -1 \\
-1 & -1 & -1 & 5
\end{pmatrix}.
$$

We now need to form the right-hand side vector of our matrix system. Our rows have the same association with the teams. So, the first row corresponds to the College of Charleston, the second to Furman, and so forth. Each entry in the vector equals $1 + 1/2(W - L)$, where W and L are the number of wins and losses by the school associated with a given row. Thinking about the first row, we are working with the College of Charleston. They won one game and lost one game, so the entry is $1 + 1/2(1 - 1) = 1$. The second entry in the vector corresponds to Furman, which won two games and lost one game. So, the entry equals $1 + 1/2(2 - 1) = 1.5$. Completing these steps enables us to write the entire matrix system:

$$\begin{pmatrix} 4 & -1 & 0 & -1 \\ -1 & 5 & -1 & -1 \\ 0 & -1 & 4 & -1 \\ -1 & -1 & -1 & 5 \end{pmatrix} \begin{pmatrix} C \\ F \\ D \\ A \end{pmatrix} = \begin{pmatrix} 1 \\ 1.5 \\ 1 \\ 0.5 \end{pmatrix},$$

where C, F, D, and A correspond to the ratings for the College of Charleston, Furman, Davidson, and Appalachian State, respectively. We can write this system as $4C - F + A = 1$, $-C + 5F - D - A = 1.5$, $-F + 4D - A = 1$, and $-C - F - D + 5A = 0.5$. Solving the linear system gives us the ratings, $C = 0.5000$, $F = 0.5833$, $D = 0.5000$, and $A = 0.4167$, which leads to the ranking (from best to worst) of Furman, a tie for the second place between Charleston and Davidson, and Appalachian State in the last place.

6.3. Get to the Point

While the Colley method doesn't take scores into account, the Massey method, as introduced in [44], does. Massey's linear system is largely the same. The only difference in the square matrix is that the diagonal elements equal only the number of games. (We don't add two.) Let's again rank our season in Figure 6.3. We'll keep the same correspondence between the rows and the teams. So, our square matrix for the Massey method becomes

$$\begin{pmatrix} 2 & -1 & 0 & -1 \\ -1 & 3 & -1 & -1 \\ 0 & -1 & 2 & -1 \\ -1 & -1 & -1 & 3 \end{pmatrix}.$$

Now, we need the right-hand vector, which uses the point differential for each game. If a team wins by 5, then the point differential is 5. If the team lost by 5, then the point differential is -5. The entry in the right-hand side vector equals the sum of the point differentials for all the games played by the associated team. Remember that each arrow in Figure 6.3 has a weight giving us the point differential of the game. So, the College of Charleston beat Furman by 10. Furman beat Davidson by 5. Davidson beat Appalachian

State by 7. So, the first entry in the vector will be $10 - 15 = -5$, since the College Charleston won by 10 and lost by 15. The second entry will be $1 + 5 - 10 = -4$, since Furman won by 1 and 5 but lost by 10. Continuing in this way gives us the linear system

$$\begin{pmatrix} 2 & -1 & 0 & -1 \\ -1 & 3 & -1 & -1 \\ 0 & -1 & 2 & -1 \\ -1 & -1 & -1 & 3 \end{pmatrix} \begin{pmatrix} C \\ F \\ D \\ A \end{pmatrix} = \begin{pmatrix} -5 \\ -4 \\ 2 \\ 7 \end{pmatrix}.$$

It turns out we can't find a unique solution to this system. There happen to be infinitely many. To fix this, we replace the last row of the square matrix with ones and the last entry in the right-hand side vector with a zero. This is stating that the sum of the ratings will equal 0. So, we'll be solving the system

$$\begin{pmatrix} 2 & -1 & 0 & -1 \\ -1 & 3 & -1 & -1 \\ 0 & -1 & 2 & -1 \\ 1 & 1 & 1 & 1 \end{pmatrix} \begin{pmatrix} C \\ F \\ D \\ A \end{pmatrix} = \begin{pmatrix} -5 \\ -4 \\ 2 \\ 0 \end{pmatrix}.$$

Solving the system gives us $C = -2.125$, $F = -1.000$, $D = 1.375$, and $A = 1.750$, which leads to the ranking (from best to worst) of Appalachian State, Davidson, Furman, and Charleston. In comparison with the results of the Colley ranking for this collection of games, we see that the Massey ranking rewards for large wins as was the case with Appalachian State over Charleston. It also breaks the tie for the second place between Charleston and Davidson. An added benefit of the Massey method is the inherent meaning of the computed ratings, which is connected to the method's derivation. The difference in the two teams' ratings predicts the point differential of future games. So, the Massey method predicts that if Charleston and Davidson were to play, Davidson would beat Charleston by $1.375 - (-2.125) = 3.5$ points.

6.4. Not Just Another Win

There seems little madness in applying these methods to create brackets for the NCAA basketball tournament. A bracket is formed by simply assuming that a higher ranked team wins. We all just learned the same method.

So, everyone would form the same bracket, which may not be desired. A simple adjustment to the system results in personalized brackets. Using the ideas introduced in [25] and detailed in [41], you simply decide on a weight for each game. Our previous methods had each game corresponding to one win and one loss. Now, we hope to make those games that are more predictive of how a team will play in March Madness worth more than those games that are not.

Suppose we decide that the second half of the season is more predictive of performance in March Madness. Let's make games in the first half worth 0.75 and games in the second half worth 1.25. The change to the linear systems is minor. Now, a game is simply counted as the weight of the game in its contribution to the linear systems. Before C and M were formed with each game counting as 1; now it is simply the associated weight. So, returning to our example of breaking a season into two parts, a game would count as 0.75 of a game in the first half of the season. As such, the total number of games becomes the total number of weighted games. The only other difference is the right-hand side of Massey. Now, the point differential in a game is a weighted point differential, which equals the product of the weight of the game and the point differential in that game. So, a game won by 6 points in the first half of the season would have a (weighted) point differential of $6(0.75) = 4.5$.

Let's return to the NCAA basketball example shown in Figure 6.3 to investigate the effect of incorporating recency into the Colley and Massey calculations. In Figure 6.4(a), the number next to an arrow indicates the day of the season in which the associated game was played. Let's take the length of the season to be seven days and weight each game such that games in the first half of the season are weighted by 1/2 and games in the second half count as a full game.

The weighted Colley method results in the linear system

$$\begin{pmatrix} 3.5 & -1 & 0 & -0.5 \\ -1 & 4 & -0.5 & -0.5 \\ 0 & -0.5 & 3.5 & -1 \\ -0.5 & -0.5 & -1 & 4 \end{pmatrix} \begin{pmatrix} C \\ F \\ D \\ A \end{pmatrix} = \begin{pmatrix} 1.25 \\ 1 \\ 1.25 \\ 0.5 \end{pmatrix}.$$

Solving the linear system, we find that $C = 0.5581$, $F = 0.5065$, $D = 0.5419$ and $A = 0.3935$, which leads to the ranking (from best to worst) of Charleston, Davidson, Furman, and Appalachian State.

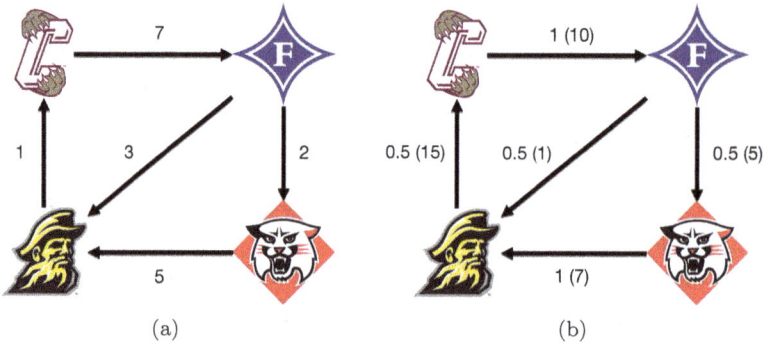

(a)   (b)

Figure 6.4:   A fictional season of games played between four NCAA basketball teams. The number next to an arrow in (a) indicates the day of the seven-day season that the game was played. The same season is given in (b), but now the arrows are labeled with the weight for each game, and the difference between the winning and losing scores is given in parentheses.

Let's look at the same series of games with the Massey method. Figure 6.4(b) displays the weight for each game (derived from its day played), and the difference between the winning and losing scores is given in parentheses. The linear system becomes

$$
\begin{pmatrix}
1.5 & -1 & 0 & -0.5 \\
-1 & 2 & -0.5 & -0.5 \\
0 & -0.5 & 1.5 & -1 \\
1 & 1 & 1 & 1
\end{pmatrix}
\begin{pmatrix}
C \\ F \\ D \\ A
\end{pmatrix}
=
\begin{pmatrix}
2.5 \\ -7 \\ 4.5 \\ 0
\end{pmatrix},
$$

resulting in the ratings $C = -0.0476$, $F = -2.8095$, $D = 2.3810$, and $A = 0.4762$. Ranking (from best to worst) gives us Davidson, Appalachian State, Charleston, and Furman. Note how the effect of Appalachian State's large win over Charleston was lessened since it happened early in the season. Now, if Charleston and Davidson play, this weighted Massey method suggests that Davidson would beat Charleston by $2.3810 - (-0.0476) = 2.4286$ points.

6.5. Rating Madness

How do these methods perform in creating brackets for March Madness in 2014? Counting every competition as one game, the Colley method correctly predicts 66.7% of the games and the Massey method has 63.5% accuracy. Note that the two methods vary in which performs best from year to year.

Using our weighting that a game in the first half of the season is worth 0.75 and that in the second half equals 1.25, Colley's performance degrades slightly to 61.9% accuracy. The Massey method jumps to 69.8% accuracy.

Keep in mind that other aspects of a game can be weighted, such as whether a team wins at home or away. Or else, one might weight a team's ability to retain winning streaks. I've used this in the class with my students. In 2010, a student produced a weighted method that outperformed 99% of over 5 million brackets in the 2010 ESPN online bracket competition. Results vary, in part because each student is making personal modeling decisions. What are yours? Can you harness some of the Madness of March with your decisions? What's your ideal weight?

Chapter 7

Pigeonholing an Athlete

Suppose that you are at a Major League Baseball game. Your friend, who isn't an avid baseball fan, is looking through binoculars at the stands for celebrities and suddenly says, "Did you know that the average person at this game is a millionaire?" You just became an average millionaire!

How could this happen? In 2018, Bill Gates' net worth was 90 billion according to Wikipedia. The largest baseball stadium is Dodgers Stadium holding 56,000. If your friend spotted Bill Gates in the stands, then the average net worth within the stadium is $90,000,000,000/56,000 > 1,000,000$. So, regardless of which stadium you were at, you became an average millionaire — as long as Gates is in attendance, of course (Figure 7.1).

Would the same be true if you (and Gates) attended a sold-out event with a crowd of 76,000 to see a United Manchester soccer match at Old Trafford stadium? What if you (and Gates) are part of a crowd of 100,000 at Michigan Stadium? These puzzles are for you to solve! You can check your answer with that in the footnote.[a]

7.1. Knowing Two Needles are in a Haystack

Looking at the world through the mathematical lens can give us insight otherwise obscured. We can consider questions that may seem unanswerable

---

[a]Old Trafford seats (even at record capacity) less than 80,000 and $90,000,000,000/80,000 > 1,000,000$. However, with Michigan Stadium, $90,000,000,000/100,000 < 1,000,000$.

Figure 7.1:   You could be an average millionaire at Dodger Stadium. Picture by Alessandro Porri.

but are solvable with math. The question below will require using a similar logic to that used in the opening of this chapter.

Question: Do two people exist such that they have exactly the same number of hairs on their heads?

Imagining a mechanism to answer this question can be hair-raising. I'm not counting every hair on even one head let alone one after another until two had the same number. Even if I counted correctly, which would be quite a feat, the two people may not have the same number of hairs at the same time since one person could lose a hair between the time I counted one person's hairs and finished counting the second person's (Figure 7.2).

Where do we start? Let's digress and share a whimsical and largely fabricated tale. One day, unknown to anyone except the small toddler himself, my son mixed white glue into my bottle of conditioner. The next morning I took a shower. About an hour later, my son sat on my lap as I explained how glue does not belong in conditioner and, for the sake of complete clarity, in shampoo bottles either. I tried and tried to comb the glue out of my hair. In the end, I managed to pull out a patch of hair that was exactly 1 square inch — no more, no less. The irresistible lure of numbers inspired me to count the recently uprooted hairs, which totaled exactly 2,000.

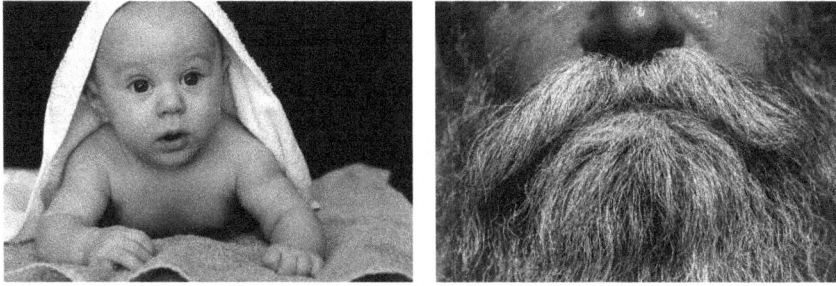

Figure 7.2: Some people clearly do not have the same number of hairs on their heads! But are there two people who do?

How can my pain educate us? In particular, it can serve as a basis for answering our question about two people existing with the same number of hairs on their heads. We want an upper bound for the number of hairs on anyone's head. We can't use the number of hairs I found on my head since you and I probably differ in our density of hair. Instead, let's imagine someone named Maxine who has more hairs than anyone on the Earth. First, we'll give her a huge head. To simplify things, she'll have a spherical head that's 2 feet in diameter and completely covered with hair. So, the area of Maxine's head is $\pi(24^2)$ or 1,809 square inches, which we'll round up to 2,000 square inches. Remember, my head had 2,000 hairs in a square inch. Using that figure, Maxine would have $(2,000)^2$ or 4 million hairs on her head!

Does Maxine have more hairs on her head than anyone on the Earth? We don't know. While Maxine's head is larger than anyone, the number of hairs may not be more. This is simply the number of hairs found on a gluey patch of uprooted hair from my head. Let's max out the number of hairs on Maxine and give her 20,000 hairs per square inch. So, her head has a surface area of 2,000 square inches and each square inch has 20,000, giving her 40 million hairs on her head. Why is this important? Maxine, if she existed, would now have the maximum number of hairs of anyone on the planet.

Approximately 7 billion people live on the Earth. Compare this number to the number of hairs on Maxine's head. If 7 billion people all have 40 million or less hairs on their heads, do two people exist with exactly the same number of hairs? Yes. They must. Imagine if they didn't. Then, line up 7 billion people and every person must have a unique number of hairs. Let's assign to each person the number of hairs on their head. So, the first person has 1 hair, the second 2 hairs. We are fine until we get beyond 40 million. We can't have someone with 1 billion or even 50 million hairs on their heads. There aren't that many options. So, somewhere in that line of 7 billion people, we must begin to repeat numbers. In doing so, two people must have the same numbers of hairs on their heads. Who are they? That's a different question and one that we will not be answered by means of mathematics. In fact, we simply aren't even going to try to answer that one.

We just used the pigeonhole principle. The statement, although simple, can be quite powerful. The principle simply states that if $n > m$ and n items are put into m containers, then at least one container must contain more than one item. For example, if we have $n = 10$ pigeons and $m = 9$ pigeonholes, as shown in Figure 7.3, then at least one hole must contain more than one pigeon.

Does someone actually have 4 million hairs on their head? We used a model from a whimsical tale. Turning more scientific, Rainer Flindt's book *Amazing Numbers in Biology* states that humans range from 90,000 to 150,000 hairs on their heads with red heads having 90,000 and blondes having 150,000 [57]. Given these numbers, no human should have more than 500,000 hairs on their head. So, we'll take this as our upper bound. There are just over 24 million people in Shanghai. So, there are at least 2 people with exactly the same number of hairs on their head in that city.

Figure 7.3: If there are more pigeons than holes, than at at least one hole has more than one pigeon.

In fact, there are at least 48 people with the same number of hairs on their heads. Who? We don't know. And, given that someone might have a hair fall-out, who are those 48 people can change. But we know they exist!

7.2. Birthday Buddies

Let's move over to Tokyo, which has over 13 million people. Are there at least two people who were born in the same hour on the same day? To solve this, let's think about an upper bound for the number of hours since someone was born. No one has lived to be 150. There are under 1,400,000 hours in 150 years. So, indeed, there are at least nine people in Tokyo who were born during the same hour on the same day.

Let's try another. Suppose you are at Michigan Stadium, known as the Big House. It's one of the larger stadiums of the world and is pictured in Figure 7.4. Its record attendance was under 120,000. Let's assume 120,000 fans pack themselves into the stadium for a really huge game. The athletic contest is decided by a huge play in the last second. Someone exclaims, "That about scared me to death!" Being a mathematician and seeing the size of the crowd, you wonder, "Are we guaranteed that at least 2 people in

Figure 7.4: Michigan Stadium is one of the largest stadiums in the world.

that crowd will die on the same day?" Given our work, you solve this one! Check the footnote to compare your answer.[b]

7.3. Pigeonholed Record

Let's see another example of the pigeonhole principle in sports. Since 2002, 32 teams play in the NFL within the NFC and AFC conferences. Pick either conference. In any year except 2007, 2008, and 2017, you'll find that at least two teams have the same record. I haven't even checked their records. I simply know it is true. Why? First, let's see why we had to exclude a few years. In 2007, the New England Patriots went 16-0 in the regular season. In 2008 and 2017, the Detroit Lions and Cleveland Browns had winless seasons. We can pick either conference. Let's pick the NFC. There are 16 teams. Each team plays 16 games. How many losses are possible during the regular season? You can lose 0, 1, 2, 3 to all 16 games. That's 17 numbers. But we are guaranteed that no team went winless or undefeated, which leaves us with 15 options for 16 teams.

In the NFL, some games are played between teams in the AFC and NFC. This same idea applies to any round-robin tournament, which is where each team plays with every other team exactly once. The pigeonhole principle again guarantees that there must be two teams with the same number of wins if no team was winless. Can you see why? Feel free to try it yourself before reading on.

Suppose there are six teams in the round-robin tournament. Then, every team plays five games. If no team was winless, then the teams could have won one, two, three, four, or five games, giving us five options for six teams. If you are ranking by winning percentage, you'll always have a tie if no team was winless.

This section outlines an important feature of mathematics. The pigeonhole principle is simple, almost obvious. Yet, it is extremely powerful. This gives us insight. We are told not to judge a book by its cover. It's important that we also do not judge the power of a math theorem (or principle) by how simply it can be stated.

---

[b]120,000 days is over 320 years. No one will live that long, so at least two people in that crowd will die on the same day.

Chapter 8

Being Average

We often think in averages. Are you the average height for your age? What's the average test grade in a class? Another option to reflect a middle value is the median, which separates the higher half from the lower half of a dataset. For example, for the values $1, 3, 5, 7, 9$, the median is 5, which is the third smallest and third largest of the values. If there are an even number of values in the set, then the median equals the average of the two middle values. For example, for the values $1, 3, 5, 7, 9, 11$, the median is $(5 + 7)/2 = 6$. The median is commonly used in analytics although in sports we often talk about averages such as batting averages in baseball or average yards per carry by a running back in football.

8.1. Mean About Barry Sanders

Barry Sanders is among the leaders of NFL rushing with 15,269 yards over his career with the Detroit Lions from 1989 to 1998. In 1996, the team started the season with four wins in their first six games. They then lost nine of their final ten games. The Lions rebounded in 1997, finishing 9–7 and qualifying for the playoffs for the fifth time in seven seasons. The team lost its Wild Card Playoff game with the Buccaneers. Sanders had only 65 yards in 18 carries, equating to an average of 3.6 yards per carry. If only he played more like earlier in the season, when the Lions beat the Buccaneers 27 to 9. In that game, Sanders had 215 yards on 24 carries, equating to an impressive 9.0 yards per carry. What an incredible difference!

In that regular season win, Sanders rushed for two touchdowns. In the playoff loss, Sanders scored no touchdowns. Yet, there is more to this story.

In the 27 to 9 regular season win, Sanders' rushing touchdowns came from an 80 yard and 82 yard dash down the field. These are very long runs, and there were two of them. How much did these affect the average? If you find the median for the games, Sanders had a median of 3.5 yards per carry for the playoff game. For the 27-9 win against the Bucs, Sanders' median was 1.5 yards per carry. The average helps us to see that, overall, Sanders had a big game but comparing the average to the median demonstrates that the mean is likely skewed by a few big rushing plays, like he had. It's always good to ask for both the mean and the median to see if there are outliers pulling the average one way or another.

8.2. Million-Dollar Pay

One sports number that interests athletes is pay. Some of the world's best athletes receive jaw-dropping salaries. How accurate is our sense of the size of these figures? Averages can help us to reframe salaries to possibly give a better sense of their magnitude.

According to the Forbes list of the highest paid athletes in the world for 2018, Lionel Messi earned $111 million from salary and endorsements [3]. Cristiano Ronaldo earned $108 million. Each player generally plays about 35 matches per year, resulting in approximately 3,150 minutes or 189,000 seconds of regulation play. So, Ronaldo and Messi earned an average of over $570 per second or over $34,000 per minute.

However, let's transition to Floyd Mayweather Jr., the highest paid athlete of 2018. Mayweather earned $285 million for a boxing match that lasted 36 minutes against Conor McGregor. The purse was $275 million with $10 million in endorsements. So, Mayweather earned approximately $132,000 per second. As mentioned in Chapter 2, it takes 0.4 seconds to blink your eye. In the blink of an eye, Mayweather earned over $52,000 dollars. He earned a million dollars every 8 seconds. Note, that 8 seconds is about as long as a well-executed, IndyCar racing pit stop.

It turns out Mayweather's blink isn't enough to pay for a Super Bowl ad in 2020. According to CNBC, a 30-second ad in Super Bowl LIV was $5.6 million [30]. That's just over $186,000 a second or $46,000 in the blink of an eye. But, keep in mind, the average person blinks 15–20 times per minute. If we assume blinking 15 times per minute, then our eyes are closed for about 10% of a Super Bowl ad [60]. So, people are missing over half a million dollars of visual content during a 30-second Super Bowl ad!

But we don't sense that we are missing anything when we blink. It's likely you haven't reversed a programmed commercial to see what you missed in 0.4 of a second. So, how much really happens in the blink of an eye? A 99 mph fastball reaches the plate in 395 milliseconds. This is so fast that a baseball is invisible to the batter when it is within 15 feet of home plate, since the images from the last 15 feet can't be processed by the brain in time! Even more, the batter should not blink when the ball is released for the entire flight is over in that blink of the eye.

Let's consider Michael Phelps winning his historic seventh gold medal in the 2008 Beijing Summer Olympics. After a close race in which Milorad Cavic dominated the first half of the 100-m butterfly event, Phelps reached legendary status by touching the wall first in Olympic record time of 50:58, only a hundredth of a second ahead of Cavic. Phelps' lead was only a quarter of a blink.

Computing averages can give us the sense of a number. How much is 285 million dollars? How small is a hundredth of a second? In analytics, sometimes it isn't just a matter of computing a number but also giving a sense of its size.

8.3. Eyeing a Black and White Image

Now, let's use averages to create mathematical art. To get your creative juices flowing, look at the mathematical art in Figure 8.1 created by Dr. Robert Bosch, a professor at Oberlin College. The mosaic of Abraham Lincoln uses 12 complete sets of dominos, which means every single domino in 12 sets of dominos is used in the piece. Bosch's work uses optimization methods for his artistry [7].

Let's create a piece by hand, with black and white squares rather than dominos. On p. 85, there are two grids shown: one grid has a portrait of Abraham Lincoln behind it, and the other grid is blank. Color a square black in the blank grid if the corresponding square in the grid containing Lincoln has an overall color (or average color) that seems closer to being black than white. Else, leave the square white. In a sense, you are creating your own low-resolution image of Lincoln with very large pixels.

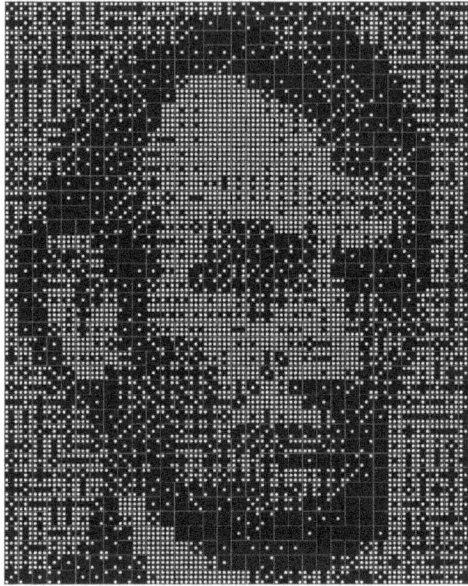

Figure 8.1: A domino mosaic by Dr. Robert Bosch. Note how dominos are placed horizontally or vertically.

Now, let's create mosaics with mathematics. On the previous page, you used your eyes to judge the color. Now, we'll develop mathematical steps to decide whether to color a square black or white.

First, in a computer, pictures are stored in tables. If the image is grayscale, the numbers in the corresponding table are between 0 and 255 and represent the intensity of the associated pixel in the image, where 0 is black and 255 is white. The table in Figure 8.2(a) gives grayscale values for an image with four rows and three columns of pixels. The pixel in the upper lefthand corner of such an image would have the grayscale value of 177, which would be gray. The corresponding image (with very large pixels) is shown in Figure 8.2(b).

As another example, consider the table with 10 rows and 10 columns in Figure 8.3(a). If the values are turned into grayscale pixels, they become the image, with very large pixels, in Figure 8.3(b). These pixels appear as part of the lower part of Lincoln's right eye in Figure 8.4(a).

Similar to how you colored the blank grid earlier in this chapter, if a grayscale value is less than 128, then the corresponding pixel is closer to

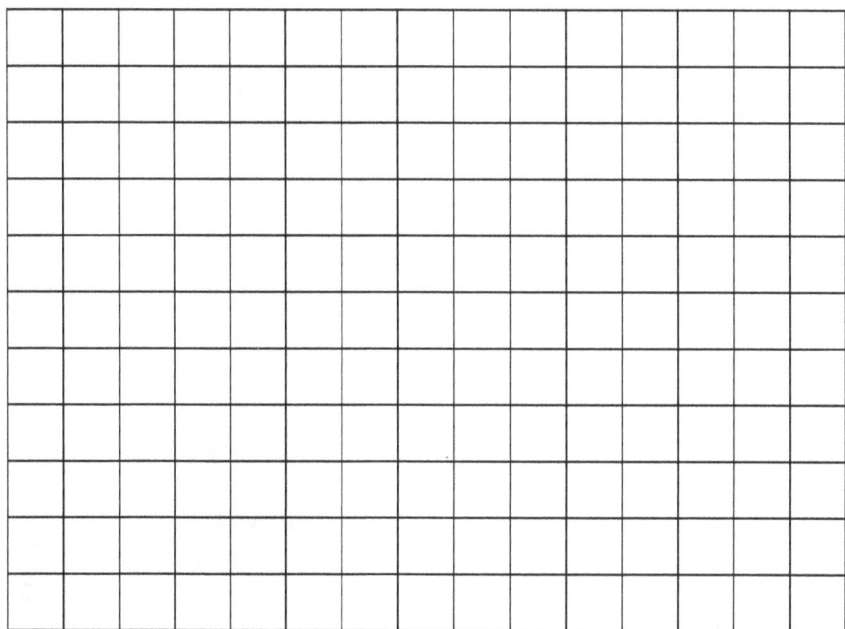

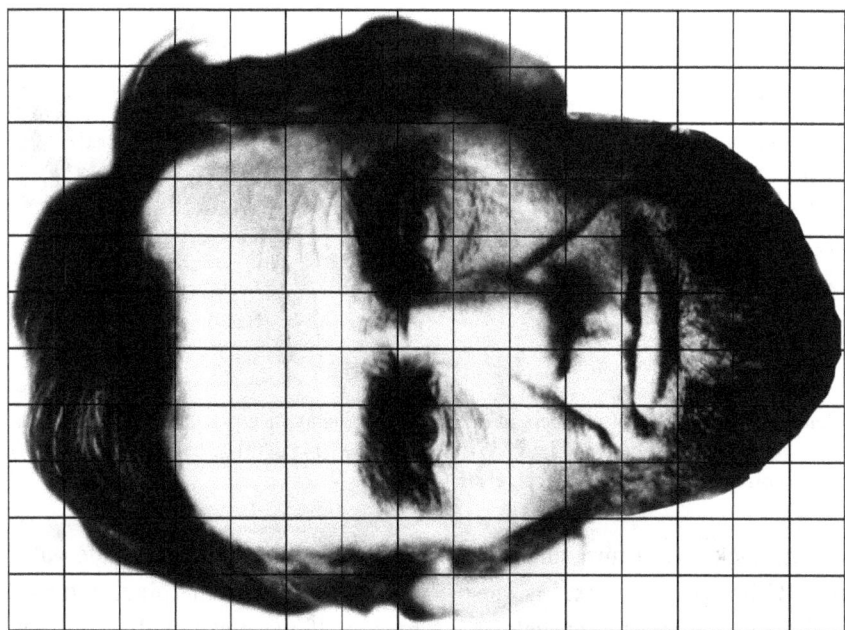

177	112	47
81	97	125
243	195	114
8	203	165

(a)

(b)

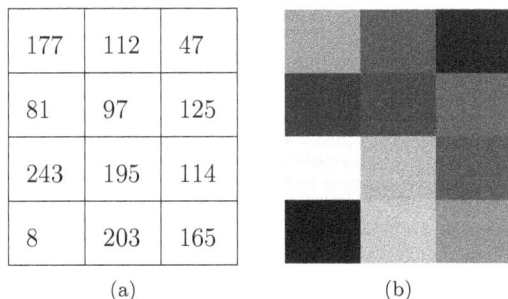

Figure 8.2:   Two 3 × 3 images of (very large) pixels.

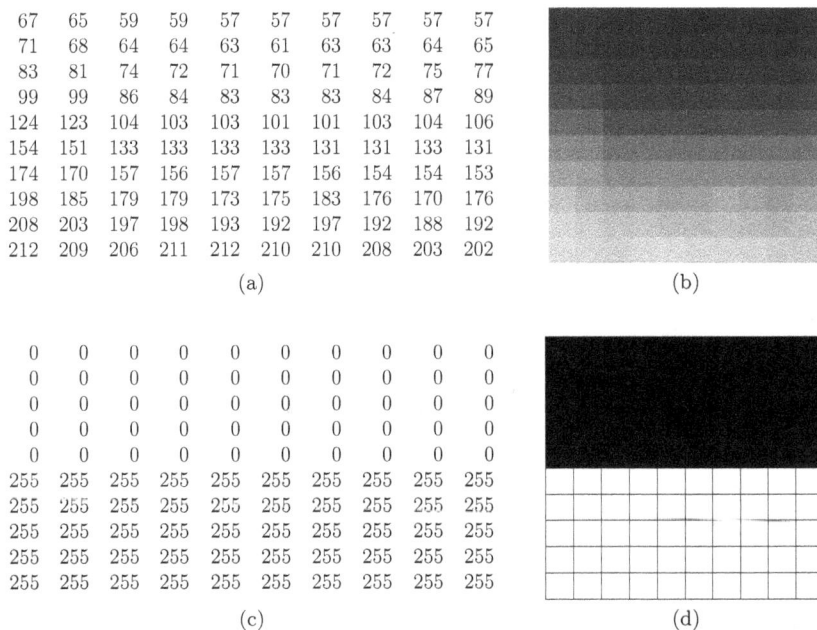

67	65	59	59	57	57	57	57	57	57
71	68	64	64	63	61	63	63	64	65
83	81	74	72	71	70	71	72	75	77
99	99	86	84	83	83	83	84	87	89
124	123	104	103	103	101	101	103	104	106
154	151	133	133	133	133	131	131	133	131
174	170	157	156	157	157	156	154	154	153
198	185	179	179	173	175	183	176	170	176
208	203	197	198	193	192	197	192	188	192
212	209	206	211	212	210	210	208	203	202

(a)

(b)

0	0	0	0	0	0	0	0	0	0
0	0	0	0	0	0	0	0	0	0
0	0	0	0	0	0	0	0	0	0
0	0	0	0	0	0	0	0	0	0
0	0	0	0	0	0	0	0	0	0
255	255	255	255	255	255	255	255	255	255
255	255	255	255	255	255	255	255	255	255
255	255	255	255	255	255	255	255	255	255
255	255	255	255	255	255	255	255	255	255
255	255	255	255	255	255	255	255	255	255

(c)

(d)

Figure 8.3:   The table of numbers in (a) becomes the grayscale pixels in (b). Turning the values less than 128 into 0 and into 255 otherwise, the table in (a) becomes the table in (c), which correlates to the grayscale pixels in (d).

being black. We'll turn such pixels black by setting their grayscale value to 0. Otherwise, the pixel is closer to being white and the pixel is turned white by setting its grayscale value to 255. If we make such a replacement, the grayscale values in the table of numbers in Figure 8.3(a) become

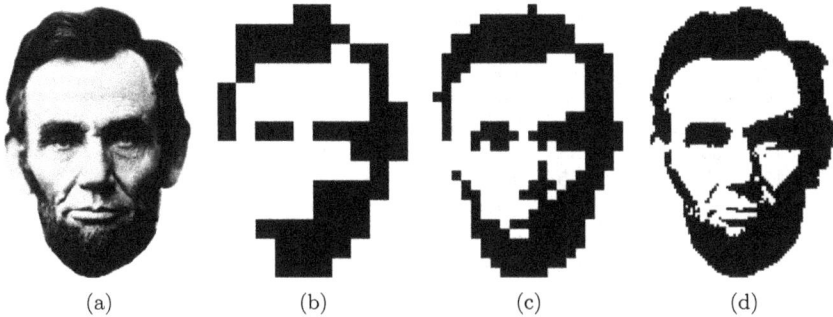

Figure 8.4: Creating approximations of the portrait of Lincoln in (a) using 10, 20, and 80 columns of solid squares as seen in the images in (b), (c) and (d).

the grayscale values in the table in Figure 8.3(c). Graphically, this corresponds to the pixels in Figure 8.3(b) being turned black and white as seen in Figure 8.3(d).

Now, rather than working with single pixel values, we'll compute the average pixel value over a square block. Similar to our previous steps, if the average is less than 128, we'll color the block of pixels in black, else the block is colored white. For example, if you take the average of the 10×10 pixels in Figure 8.3(a), you find an average pixel value of 127.2. So, such a block of 100 pixels would be replaced with black pixels. These mathematical steps will color an image like that of Lincoln.

We'll decide on the size of the blocks by determining how many blocks appear in each row. First, each row will have 10 square blocks, which for the image of Lincoln in Figure 8.4(a) becomes the image in Figure 8.4(b). If we increase to 20 columns of squares, we get the image in Figure 8.4(c) and 80 squares is seen in Figure 8.4(d).

8.4. Typewriter Artistry

We can fill the squares with something other than a color. For instance, we could place an asterisk in each square. In Figure 8.5(a), we've written the same table of Figure 8.3(a). If we replace each value less than 128 with an asterisk and with a space otherwise, we get the table in Figure 8.5(b).

If we replace blocks of pixels, just like before, but rather than using the colors black and white, we use the character "*" and a space, we get the image in Figure 8.6(a). Note how thin Lincoln's face is. To see why, look

67	65	59	59	57	57	57	57	57	57
71	68	64	64	63	61	63	63	64	65
83	81	74	72	71	70	71	72	75	77
99	99	86	84	83	83	83	84	87	89
124	123	104	103	103	101	101	103	104	106
154	151	133	133	133	133	131	131	133	131
174	170	157	156	157	157	156	154	154	153
198	185	179	179	173	175	183	176	170	176
208	203	197	198	193	192	197	192	188	192
212	209	206	211	212	210	210	208	203	202

```
*  *  *  *  *  *  *  *  *  *
*  *  *  *  *  *  *  *  *  *
*  *  *  *  *  *  *  *  *  *
*  *  *  *  *  *  *  *  *  *
*  *  *  *  *  *  *  *  *  *
```

(a)   (b)

Figure 8.5: The table of numbers in (a) becomes the block of text in (b) by replacing values less than 128 with an asterick and with a space otherwise.

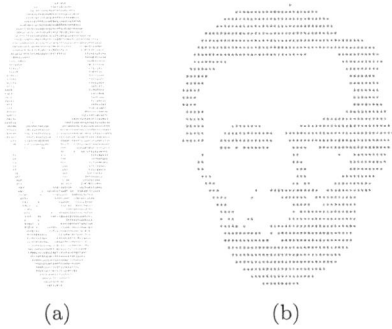

(a)   (b)

Figure 8.6: Creating an image of Lincoln with text.

carefully at the characters and how the overall height is longer than the width.

THIS FONT IS TWICE AS HIGH AS IT IS WIDE.

As written, the characters appear in a font that is twice as high as it is wide. So, we've essentially stretched our square grid by drawing the squares with a rectangular font. To correct this, we will use rectangular regions of pixels that are twice as high as they are wide.

So, rather than looking pixel value by pixel value, we look at blocks of pixels two rows by one column. So, our 10 × 10 grid gets broken into the layout that follows.

67	65	59	59	57	57	57	57	57	57
71	68	64	64	63	61	63	63	64	65
83	81	74	72	71	70	71	72	75	77
99	99	86	84	83	83	83	84	87	89
124	123	104	103	103	101	101	103	104	106
154	151	133	133	133	133	131	131	133	131
174	170	157	156	157	157	156	154	154	153
198	185	179	179	173	175	183	176	170	176
208	203	197	198	193	192	197	192	188	192
212	209	206	211	212	210	210	208	203	202

Then, we find the average of the pixels in each rectangular block. So, the upper left rectangular grid contains pixels of values 67 and 71, so the average pixel value is $(67 + 71)/2 = 69$. Continuing this process for our 10×10 grid, forms a 5×10 grid of average pixel values.

You can find the values yourself, or try a few and compare them to the 5×10 grid that follows:

69.0	66.5	61.5	61.5	60.0	59.0	60.0	60.0	60.5	61.0
91.0	90.0	80.0	78.0	77.0	76.5	77.0	78.0	81.0	83.0
139.0	137.0	118.5	118.0	118.0	117.0	116.0	117.0	118.5	118.5
186.0	177.5	168.0	167.5	165.0	166.0	169.5	165.0	162.0	164.5
210.0	206.0	201.5	204.5	202.5	201.0	203.5	200.0	195.5	197.0

We then replace each average value with an asterisk if the value is less than 128 and a space otherwise. This creates the following grid:

*	*	*	*	*	*	*	*	*	*
*	*	*	*	*	*	*	*	*	*
			*	*	*	*	*	*	*

Performing these steps on the entire portrait of Lincoln, we get the image in Figure 8.6(b).

We can also vary the character that's written. In Figure 8.7, we write out (removing spaces between words) the Gettysburg Address.

```
                              Foursc
                    oreandsevenyearsagoourfa
                 thersbroughtforthonthiscontine
                nt,anewnation,conceivedinLiberty,
              anddedicatedtothepropositionthatallm
            enarecreatedequal.Nowweareengagedinagreatcivi
          lwar,testingwhetherthatnation,oranynationsoconce
        ivedandsodedicated,canlongendu        re.Wearemeton
      agreatbattl       e        fie          ldofthatwar.
      Wehavec                                   ometodedi
    cateapo                                      rtionofth
    atfiel                                        d,asafin
    alrest                                      i  ngplacef
    orthos                                        ewhoherega
    vethei                                        rlivesthatth
    atnat                                         ionmightlive.I
    tisalt                                       ogetherfittingandpr
    operth      atweshouldo        this.But,inalargersense,wec
    annot      dedicatewecanno     tconsecratewecannothallowthis
  ground.T    hebravemen,liv      inganddead,whostruggledhere,h
  av  econ    secratedit,fara      boveourpoorpowertoaddordetr
      act          .Th            eworldwilllittlenote,norlo
     ngr                          emememberwhatw  esayhere,bu
      tit                         canne           verforgetwh
      att                         heyd           idhere.Itisf
     orust                        helivi      ng,rather,tob
    ededic          at edheretotheunfinishedworkwhic
    htheywh    of       oughtherehavethusfarson
      oblyad va        n ced.Itisratherforus
      tobehe         r    ededicatedtothegre
    attask rem       ainingbeforeusthatfromt
    hesehono    reddeadwetakeincreaseddevo
    tiontothatcauseforwhichtheygavethel
    astfullmeasureofdevotionthatwehere
    highlyresolvethatthesedeadshallno
    thavediedinvainthatthisnation,u
      nderGod,shallhaveanewbirtho
      ffreedomandthatgovernme
        ntofthepeople,byth
          epeople,for
          thepeople,
      shall not perish from the earth.
      Abraham Lincoln, November 19, 1863
```

Figure 8.7:   Creating an image of Lincoln with text from the Gettysburg address.

8.5. Colorful Math Mosaics

Let's return to making square blocks of pixels in our image. Rather than using two colors, black and white, we can use three colors, black, gray and white. To do this, we color any square block that has an average pixel value of less than 85 with black. If the average pixel value of a square block is greater than or equal to 85 or less than 170, then we color it gray. Else, the square block is colored white. Using this on our portrait of Lincoln with 20, 40, and 80 blocks, we get the images in Figure 8.8.

If we use the same logic and increase to six colors, we can use six shades of gray to color the image. However, let's instead replace each square with the image of one side of a six-sided die. The region that would be replaced

Figure 8.8: Creating approximations of the portrait of Lincoln using 10, 20, and 80 columns of solid squares with three colors.

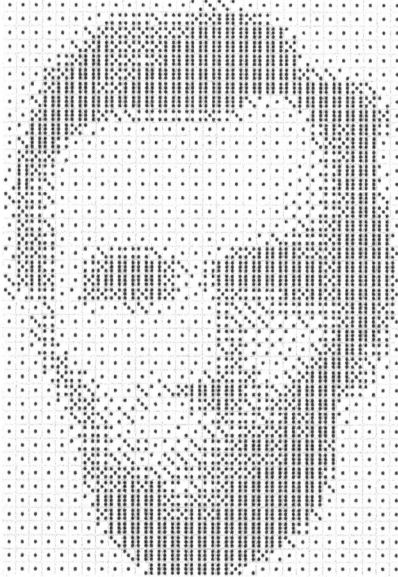

Figure 8.9: Creating approximations of the portrait of Lincoln using dice.

with white will now be replaced with the side of the die with one dot. The next lightest region will be replaced with the side of the die with two dots. The darkest region will be replaced with the side of the die with six dots. When we do this, our portrait of Lincoln becomes the image in Figure 8.9.

Is this how Dr. Bosch makes his domino mosaics? Like our work, he finds the average color over blocks of pixels. But, in his work, he uses every domino. When we used dice, we placed the side with one dot as many times as needed. Bosch only has so many of each type of domino. So, he wants to place every domino such that the resulting image is as close to the underlying portrait as possible. This is how the problem becomes an optimization problem. Create the optimal image subject to using every domino that we have. Bosch creates mathematical formulas and then inputs the equations into a computer, which can solve the problem and output a guide on where each domino is placed and whether it should be placed horizontally or vertically.

Chapter 9

Games of Chance

Let's play X Games of chance. In ancient times, rolling dice, flipping a coin, or shuffling playing cards served to generate random results. In this chapter, we will look at some methods that are based on random numbers. Such "games of chance" can lead to profound insights and realistic simulations in areas such as nuclear physics and computer graphics. To see how a large stack of random numbers can be used mathematically, we'll first estimate the unlikelihood of sporting events and then compute digits of π and in later chapters move to other applications.

9.1. Messi Calculation

Let's begin with soccer and a feat of Lionel Messi. In the fall of 2018, Messi had scored 42 direct free kicks in his career. Most notably, he scored 8 of those 42 up to that point in 2018 (Figure 9.1). Now, Messi scored about 12.5% of his direct kicks, and took about 35–40 free kicks per year. Therefore, through September 2018, Messi would have taken about 30 free kicks and scored 8 of those 30, giving him a conversion rate of $8/30 \approx 27\%$ for that period of time. If we assume Messi on any given free kick scores 12.5% of the time, then what's the likelihood of him scoring 8 of 30?

To estimate this, you get to kick better than Messi. You'll have a 16.6% chance of scoring on a free kick. Roll a die to decide the outcome of your kick. You score a goal if you roll a one. Roll the die 30 times and count the number of times you roll a one. Did you roll a one eight times? If so, then you've accomplished Messi's feat! Did you do it? It's possible but not all that probable. You have just over a 6% chance of having your total equal 8.

Figure 9.1: Lionel Messi in the 2014 World Cup Final. This photograph was produced by Agência Brasil, a public Brazilian news agency.

If we use Messi's 12.5% conversion rate (which would equate to getting two heads on two consecutive flips of a coin), then the chances drop to about 2%. Quite a stellar 2018 for Messi!

How did I know you only had just over a 6% chance of getting 8 ones in 30 rolls of a die? You can look at the goals as successes in what are called Bernoulli trials. If you do, then you can compute

$$\binom{30}{8}(.166)^8(1-.166)^{30-8} \approx .062, \quad \text{where} \quad \binom{30}{8} = \frac{30!}{(8!)(22!)}.$$

Another way is to use computation. Pick a random number between 0 and 1; let's call it X, where every number is equally likely. If you want to simulate rolling a die, then $X < 0.166$ means you scored on your free kick. You do this 100,000 or 1,000,000 times and see what proportion of the time you made the kick. You would find this number to be about 0.06.

9.2. Flipped Out Calculation

Let's look at some other events and analyze their probability by flipping a coin. We'll begin by analyzing a coin flip. In the NFL, the team that begins the game kicking the ball is decided by a coin flip. The team captains walk to center field. A referee flips the coin and the team that wins the coin flip chooses whether to begin the game kicking (or receiving) the ball. In 2015, the New England Patriots had won the coin toss 19 of the last 25 times [65]. Give it a try. Flip a coin 25 times. Can you get 19 or more heads? It's unlikely you'll get at least 19. You only have three-quarters of 1%. Now, 25 games is about a season and a half of games. The Patriots alone have been around for over 40 seasons. When you combine all the teams over all their games, it isn't surprising that this has occurred some time.

To see that something can be improbable but still likely to happen eventually, let's consider Steph Curry's performance on February 27, 2013 against the New York Knicks in Madison Square Garden. Curry scored 54 points making 11 of 13 on three-pointers and 7 for 7 from the free throw line. Curry also led the team in rebounds and assists. He made 23 points in the second quarter alone. The performance put Curry in Hall of Fame company. Only seven players in NBA history have scored more at Madison Square Garden — Wilt Chamberlain, Elgin Baylor, Kobe Bryant, Bernard King, Rick Barry, Richie Guerin and Michael Jordan.

That night, Steph Curry missed only 2 of 13 three-point attempts which is 85%. To put this in perspective, only 6 NBA Hall of Famers have a higher career shooting percentage than 85.7% from the free throw line. So, how probable was this event? To get a sense, you're going to get to shoot better than Steph Curry. Curry was shooting threes at 45.3% that year. You get to shoot 50% by flipping a coin. Flip a coin 13 times and see if you get 11 or more heads. It may happen but not frequently. You have just over 1% chance. One way to estimate the probability is to flip a coin 13 times and see if you get 11 or more. Then, do it again and again and again. If X is the total number of times you flipped a coin 13 times and Y is how many of those times you got heads 11 or more times, then you'd find that Y/X is about 0.01 if you take X to be large, like 10,000.

Let's consider Curry's performance in another way. He shot about 600 three-pointers during the regular season and just over 100 in the playoffs. Imagine flipping an unfair coin that comes up heads 45.3% of the time. If we flip the coin 700 times, how often will we see a string of 11 heads in 13 consecutive flips? This happens about 60% of the time! Even so, on any given night, the probability of making 11 of 13 on three-pointers is very very low.

9.3. Math Pub Darts

Let's take a closer look at using repeated experiments for mathematical calculation. We'll imagine playing an especially random game in the Math Pub, which is a place where all things mathematical but not necessarily physical can occur. We will play on a circular dartboard with unit radius inscribed on a square board. We play together rather than against each other and need to play in the Math Pub so our throws adhere to the assumptions:

- Every dart will hit the square board.
- It is equally likely that every position on the square board will be hit by a dart on any given throw.

If you play darts, this may seem like an odd combination of skills. We assume the ability to hit the dartboard on every throw and also that every throw is as likely to land in the upper lefthand corner as it is in the center of the board. Remember the goal of our game — creating a close estimate to π. These assumptions are crafted to ensure our success (Figure 9.2)

We win by keeping count of the number of throws that land in the circle and the total number of tosses. Note that the proportion of darts that land in

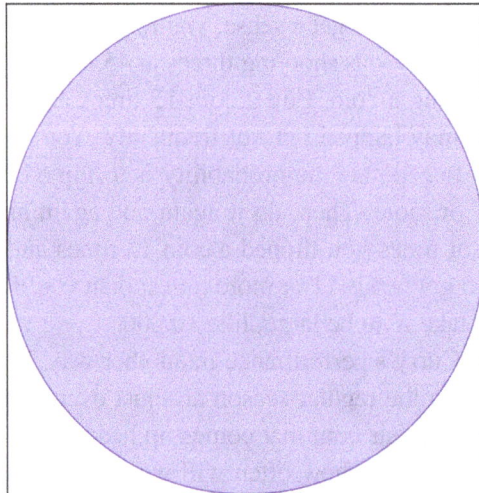

Figure 9.2:   A dartboard in the Math Pub for estimating π.

the circle to the total number of darts tossed will approach the fraction:

$$\frac{\text{area of the circle}}{\text{area of the square box}} = \frac{\pi}{4}$$

as we toss more and more darts. As we play, the fraction

$$4 \cdot \left(\frac{\text{number of darts that landed in the circle}}{\text{total number of darts thrown}} \right)$$

will approach the value of π.

Now, let's alter the game a bit so you can easily produce the necessary random numbers. We will play on a quarter of the dartboard. Rather than choosing random real numbers for our coordinates on the dartboard, we will choose integers between 2 and 12 by summing the numbers that appear from a roll of a pair of dice.

Follow these steps:

1. Roll the pair of dice.
2. Identify the result in the game board in Figure 9.3. If the result is inside the circle, then your roll "landed in" the circle. For instance, if you roll

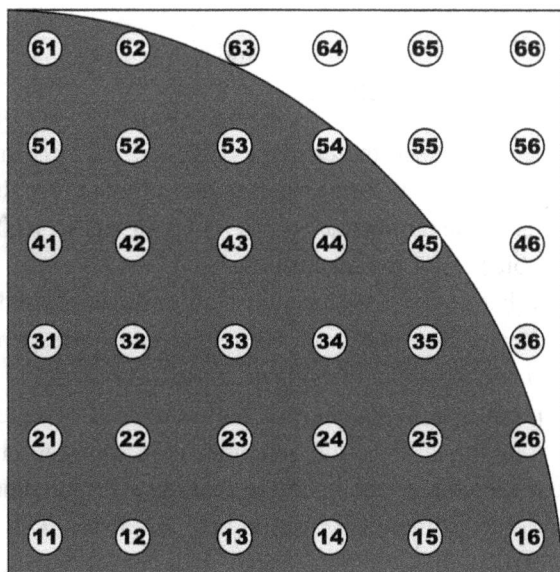

Figure 9.3: Game board for estimating π with a pair of dice.

Table 9.1: Approximations to π after rolling a pair of dice 10,000 or 100,000 times.

Number of rolls	Estimate to π
10,000	3.1056
10,000	3.1056
10,000	3.1032
10,0000	3.1050
100,000	3.1159
100,000	3.1014

a 6 and a 1, then look for 61 in the game board. We see that such a roll would land inside the circle.

3. Keep track of the number of rolls that land in the circle and the total number of rolls.

Let one experiment consist of rolling the pair of dice 10 times and then compute the result:

$$4 \cdot \left(\frac{\text{number of rolls that landed in the circle}}{\text{total number of rolls}} \right).$$

A few experiments produce the estimates 3.6, 3.2, 3.2, 3.6, and 2.8. With only 10 tosses, we are not winning our game of producing a close estimate of π. Let's increase the number of rolls per experiment to 10,000 and then 100,000; see Table 9.1 for the results.

That's much better. Still, the results seem to hover around 3.10. Look carefully at the board. Note that 28 of the 36 outcomes correspond to landing in the circle. So, we are actually estimating $4(28/36) = 3.1111$.

Could we produce a better approximation to π if we alter our board so that more or less than 28 of the outcomes correspond to landing in the circle? Suppose we change the board so that only 27 outcomes landed in the circle. Now, our estimate to π would be $4(27/36) = 3$. If we alter the board so that 29 outcomes are in the circle, then our estimate to π becomes $4(29/36) = 3.2222$. Our original choice was indeed the best and leads to the closest approximation.

Admittedly, there are much faster ways to compute π as this approach would not be useful to find, say, 10,000 digits of π.

9.4. Klutzy Math

Can't find a Math Pub or don't like darts? Let's look at another way to calculate with randomness. In 1773, Georges-Louis Leclerc, Comte de Buffon, described a game in which two gamblers drop a loaf of French bread on a wide-board floor. They bet on whether or not the loaf falls across a crack in the floor. Buffon asked what length of the loaf of bread, relative to the width of the floorboards, would lead to a fair game, which he answered in 1777.

Let's create a version of this experiment seen in Figure 9.4. Here's how to create it yourself.

1. Find a needle (or even a toothpick) and measure its length.
2. Mark a piece of paper with 2 lines equally spaced at a distance equal to the length of your needle. You can see this in Figure 9.4.
3. Drop your needle on the paper.
4. Keep track of the total number of times you drop the needle and the number of times a dropped needle crosses a line.
5. Compute $C = \dfrac{\text{total number of crossings}}{\text{total number of drops}}$.

As best you can, drop the needle with a random rotation and distance from the lines. I dropped my needle five times and got three crossings. For my Buffon Experiment, $C = \frac{3}{5} = 0.6$. What did you get? I did it again and got four crossings. Given this is random, let's try it many more times

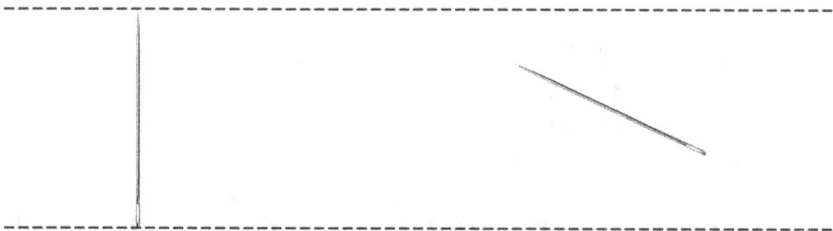

Figure 9.4: Buffon's Needle experiment can be analyzed mathematically. Let y denote the distance from the southern end of the needle to the upper line and θ is the angle of the needle to the horizontal. So, $d = L \sin \theta$.

like we did for throwing darts. I already did it many more times — in fact, many, many more times. Here is what I got!

Total drops	(Total crossings)/(Total drops)
100	0.66000
1000	0.64300
10000	0.63900
100000	0.63391
1000000	0.63676
10000000	0.63680

And? Turns out that we were estimating π. How? Rather than computing $C = $ (total number of crossings)/(total number of drops), let's compute

$$P = 2 \cdot \left(\frac{\text{total number of drops}}{\text{total number of crossings}} \right).$$

I'll keep the results of my previous tosses but now compute P rather than C. Here is what I get.

Total drops	2(Total drops)/(Total crossings)
100	3.03030
1000	3.11042
10000	3.12989
100000	3.15502
1000000	3.14088
10000000	3.14071

This can seem impossible. Math helps uncover the process. Let's call the length of your needle L. So, our lines are a distance of L away. We'll measure the distance from the southern end of the needle to the upper line and call this distance y. We'll let θ be the angle of the needle to the horizontal.

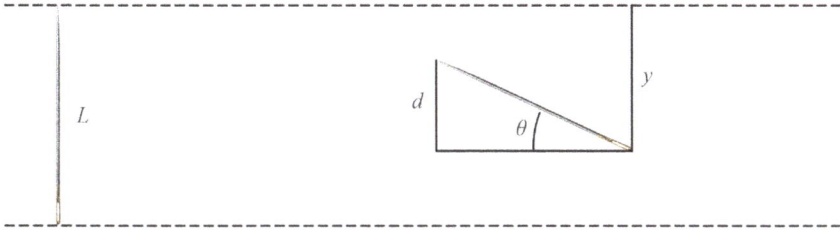

Figure 9.5: Buffon's needle experiment randomly drops a needle of length L between two lines a distance L apart.

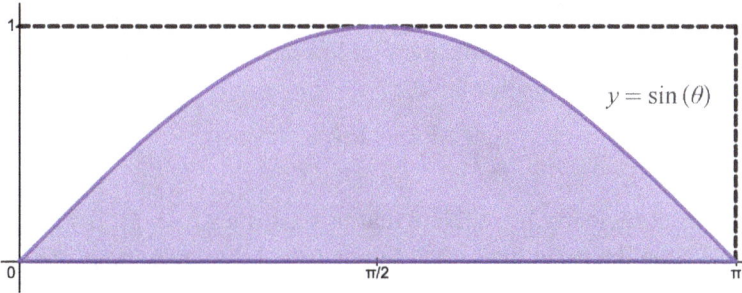

Figure 9.6: Plot of $\sin\theta$ aids in analysis of Buffon's needle.

To see this visually, see Figure 9.5. Note that the needle is crossing a line if

$$y \le L\sin\theta.$$

Let's simplify things by taking $L = 1$, so there is a crossing if $y \le \sin\theta$. Consider the plot of $\sin\theta$ shown in Figure 9.6. We can view this like the earlier dartboard. If you randomly throw a dart at this board, it would land at a point (θ, y), which corresponds to dropping the needle at an angle (θ) and a distance y from the top line. If the dart lands in the shaded region (below $\sin\theta$), then we have a crossing. What's the probability of getting a dart, thrown at random, to hit the shaded portion of this board? It's the ratio of the area of the shaded region to the area of the rectangle. The area of the rectangle is π. The area of the shaded region is

$$\int_0^\pi \sin\theta \, d\theta = 2,$$

which implies that the probability of a hit is

$$p = \frac{1}{\left(\frac{\pi}{2}\right)} = \frac{2}{\pi}.$$

Hence, an estimate to π is

$$\pi \approx 2 \cdot \left(\frac{\text{total number of drops}}{\text{total number of hits}} \right).$$

Dropped a pencil on a floor with lines? Say, you're estimating π. But what if the distance between the lines is greater than the length of your pencil? Using a similar analysis to that we just did, you can derive that if you drop a needle of length L a total of n times and find that the needle crosses a line c times where the parallel lines are a distance of D apart, then the following approximation holds:

$$\pi \approx \frac{2nL}{Dc},$$

where n is sufficiently large. Note that this assumes $L < D$. If your pencil is longer than the ruled lines, then things get a bit more complicated as

$$\pi \approx \frac{n}{Dc} \left\{ D \left[\pi - 2\sin^{-1}\left(\frac{d}{l}\right) \right] + 2L \left(1 - \sqrt{1 - \frac{D^2}{L^2}} \right) \right\}.$$

You can also use this experiment, called Buffon's needle, to pass the time. This is what Captain O. C. Fox did as he recovered in a military hospital

Table 9.2: Super Bowls between 2004 and 2013 and Madden NFL predictions.

Year	Super Bowl	Madden's winner	Madden's loser	Correct prediction
2004	XXXVIII	New England Patriots	Carolina Panthers	Yes
2005	XXXIX	New England Patriots	Philadelphia Eagles	Yes
2006	XL	Pittsburgh Steelers	Seattle Seahawks	Yes
2007	XLI	Indianapolis Colts	Chicago Bears	Yes
2008	XLII	New England Patriots	New York Giants	No
2009	XLIII	Pittsburgh Steelers	Arizona Cardinals	Yes
2010	XLIV	New Orleans Saints	Indianapolis Colts	Yes
2011	XLV	Pittsburgh Steelers	Green Bay Packers	No
2012	XLVI	New York Giants	New England Patriots	Yes
2013	XLVII	Baltimore Ravens	San Francisco 49ers	Yes

from a severe wound in the American Civil War. Fox threw a fine steel wire onto a wooden surface ruled with equidistant parallel lines. Fox got 3.1780 for π in his first experiments. Keep in mind that the experiment assumes randomness of placement, which involves the rotation of the needle. So, Fox, wanting to decrease any bias in his throws, added a slight rotatory motion before dropping the rod. Fox reported improved estimates of 3.1423 and 3.1416 with approximately 500 throws [35].

There has been some question raised about Fox's work. If he had observed one more or one fewer crossing, his estimate of π would have been less accurate. Keep in mind that getting the best possible estimate is unlikely given the presence of randomness in the experiment.

9.5. Super Maddening Predictions

Simulations can seem like a game. In fact, they may even be a video game. The Madden NFL video game franchise is one of the most popular and best-selling sports games. The game can simulate the Super Bowl and makes a prediction in terms of both the winner and the final score. The game made its first prediction in 2004 and correctly predicted the champion. Madden NFL correctly predicted the Super Bowl champion 80% of the time between 2004 and 2013 as seen in Table 9.2. As such, the simulation has become one indicator of Super Bowl victories, although it can struggle. Between 2016 and 2019, Madden NFL only got 1 out of the 4 Super Bowl predictions correct. Simulation can be a powerful tool, especially when it helps in various types of realistic decision-making [39].

The methods of this chapter use randomness to help simulate our world. We've referring to the methods as simulation. They are also called Monte Carlo methods or Monte Carlo simulation.

Chapter 10

Sporting Lessons

Athletics teaches such attributes as being a good sport, dedication, and commitment. In competition, we learn to lose and may even go through losing streaks. Even so, we must play. Hank Aaron said, "My motto was always to keep swinging. Whether I was in a slump or feeling badly or having trouble off the field, the only thing to do was keep swinging." Regina Brett wrote in her book *God Never Blinks*, "No matter how you feel, get up, dress up and show up." A legendary example occurred in Game 5 of the 1997 NBA Finals when Michael Jordan walked onto the court suffering from "flu-like symptoms" throughout the game. Still, we led the Chicago Bulls to victory with a 38-point performance. Athletics is anything but constant. Greek philosopher Heraclitus is credited with the quote, "Change is the only constant in life." Take the 2008 Detroit Lions. They had the first defeated season in NFL history, going 0-16 for the season. Three years later, the team would be in the playoffs after the 10-6 season. The lessons of athletics are many.

Mathematics also has its lessons. We learn logic, accuracy, and the many concepts and ideas of the field. Yet, we learn many other lessons, too. Let's see some lessons connected to the content we've already covered in the book. Let's learn some lessons of mathematics from sports analytics.

10.1. Step into the Unknown

Mathematics can teach us to step into the unknown and explore. This is true at various levels of math — from student to researcher. Let's look at an example from my work in sports analytics. In the summer of 2010,

I received a call from a producer of the ESPN Sport Science program. Sport Science was a television program that aired short segments that examined current sporting events with analysis often involving science and mathematics. The Sport Science staff were stymied in their mathematical analysis and needed help.

Within a few minutes, I understood the mathematical problem facing the ESPN program. As we completed the call, I jotted down ideas and then performed calculations. In the end, I had a result, which was described in a report. I detailed my results so that the Sport Science staff could not only understand my work but also make any changes as the TV segment evolved during production. My work was aired as part of the segment.

This led to my phone periodically ringing with new requests for help, with varying amounts of time available to complete the analysis. Each time, I was able to offer insight that moved the production to a final piece that was aired on ESPN.

Yet, there was an important interchange in every call. In that moment lies a lesson about mathematics. When the team posed the mathematical problem, I generally did not know exactly what to do to answer their question. Said another way, I was immediately stuck. So, my first question was always the same, "When do you need to know if I think I can produce a solution?" Meeting this deadline enabled Sport Science to have confidence in my offering a solution, or time for them to find someone else.

Then, I delved into the question. I was stuck, so I inherently stepped into the unknown. I didn't expect numbers to glow and the solution to almost magically emerge from a momentary mental mist, as mathematical research is often depicted in movies. Sport Science posed difficult problems. That's why they couldn't solve them and came to me. Only careful analysis and consideration could lead to insights that would fuel my progress. I quite simply had to be willing to step into the unknown to reach a point of insight.

Stepping into the unknown is common in mathematics. We are presented with a question from a book, from someone else, or one of our own. Mathematical progress often involves getting stuck. To move toward understanding, we must step into the unknown and stand there, not knowing the path to take, and trust that our training and ingenuity will help guide us to a place of insight and action. When it comes to mathematical research, there is a simple reason we get stuck. If we knew immediately what to do, then someone would probably already have done it. Keep in mind that Sport Science only called me with problems that stumped their stellar staff.

Failed attempts offer moments for learning and gaining insight. To move in new directions, we must have patience. Further, when we create new results, we must look carefully. New insight can look odd, given that it can offer insight counter to the general assumptions. New insight can also be exciting, which won't necessarily mean it is robust or correct. However, we've also seen how subtle mistakes can be. So, a new result can also look odd (such as $0 = 1$) due to an underlying error.

Mathematics can pose difficult questions. Life itself poses difficult questions. Mathematics is a place to hone thinking skills that extend beyond the field. Throughout this book, we use mathematical thinking to study sports and delve in various areas of mathematics. The problems we have explored in the book weren't always immediately obvious at the time they arose. Keep in mind that the analysis often took several attempts to find a core insight that moved the gears toward solution. Artists sketch outlines of their final drawings before settling into their bold final strokes. The errant marks on the drawing pad led to the final drawing. Innovation comes from taking various steps that are rarely entirely linear or lead directly to the desired goal. To lean on the Einstein quote, "Anyone who has never made a mistake has never tried anything new." Be willing to get stuck. Be willing to stand not knowing how to find a solution. Such times make you a better mathematician. Only when you are willing to step into the unknown can you find a path that will take you in new directions.

10.2. What's Your Problem?

Math is full of questions. When do you use simulation and when do you search for an underlying formula? Which ratings method do you use for ranking? What approach can offer insight into an analytics question? Answering such questions effectively often requires keeping the purpose of your mathematics in mind.

At first glance data science may seem like a field where one need not define a precise question. While this may be true at times, it is generally far from the rule. Let's return to the spatial information available in the NBA mentioned in Section 3.2. As mentioned earlier, at the beginning of the

2013–14 season, the NBA installed SportVU cameras in all of its arenas. The technology captured xy-coordinates of every player on the court and all three spatial positions of the ball. Such information was recorded 25 times per second for the entire game. This led to a massive amount of data containing tremendous information on the game. With such raw data, new insights emerged like Kirk Goldsberry's heat maps, which showed the frequency of shots.

It can seem that large amounts of data will result in big insights. While NBA teams have analysts, mining such raw data offered its difficulties and challenges. Goldsberry, for example, approached the creation of heat maps from the view of his field — cartography. His question grew from his training — how could the data give spatial insight on shooting?

If one does not have a question in mind then tackling a large problem isn't like looking for a needle in a haystack. Although difficult, one can devise methods to look for a needle given the knowledge of the goal. Without a defined goal, one sifts through the haystack hoping to discover something. Yet, searching for an easily seen large gem versus a stealthy needle is quite different. As such, defining the goal can define an appropriate process.

Yet choosing a process can also be guided by careful and continued focus on the goal. I illustrate with an example from my research group. A professional sports organization posed an open problem to our group. I teach at a highly selective liberal arts college where many students desire independent research projects.

For this problem, the dataset was large and could answer many questions. Our job was to select relevant portions of the data and analyze them to come to conclusions about the posed question.

After a week, my students were moving forward but at a much slower rate than I anticipated. In the first of our two weekly meetings I learned that the students were integrating the data into a database. I thought, "Oh, of course. Why I didn't think of that? It is the natural step that everyone takes." The students noted that such work would give flexibility to future problems.

The next week we had made progress but again at a much slower rate than I originally anticipated. Such a miscalculation regarding the time needed for research isn't uncommon. We may not know the exact process of solution until the problem is solved. Yet, the difference between my expectation and the actual solution time made me pause. Why did I originally have a different expectation?

My initial impression was connected to our posed question. We needed only portions of the data. Once identified, we'd only look at those portions and have no need for the overall dataset. Said another way, we'd extract what we needed and focus our work on that smaller dataset. Creating a database, although a common and wise step with our data, was unnecessary for our problem.

We were influenced by how so many others approached the data. But those other researchers were not tackling our problem in our context. While leaning on the work of others can save time, you must also remember that your challenge may be unique. So, stay focused on your objectives. My group had inadvertently began working on methods that were ideal solutions to other questions, not ours.

In math, it can be easy to get perturbed away from your proposed problem. Sometimes, the new questions are more important than the original goal and demand attention. Other times, a blurring of focus changes one's trajectory to a different, not necessarily better or even more productive, path. Such changes happen most naturally when you or your group enters challenging stages of the work.

How do you keep focus? Ideally, you create checkpoints where you reflect on what your goals are in the project and compare those to your current processes. In a sense, you stop and ask, "What's my problem?" Once answered, you can compare this with your current path toward solution.

What if you get to a point where you can't describe your question or possibly your current steps toward solution. For that, I turn to another quote of Einstein, "If you can't explain it simply, you don't understand it well enough." Taking the time to explain what you are doing and your intended goals is worth the time. It can keep you on task, save time, and resources, and in the end help you mine through data or tasks for intended outcomes.

10.3. To Err is to Learn

As mentioned, part of working with sports data and mathematics is, frankly, failing. The tricky parts of mathematics can create a puzzle. From mistakes and misunderstandings, we learn and grow. We must accept the inevitability of erroneous directions in order to extend beyond what is known and see where we are yet to reach.

To expand on this, let's venture back to April 2016. I sat in a packed auditorium at Lenoir-Rhyne University to hear award-winning writer Anne Lamott speak as part of the college's Visiting Writers Series. She offered words of advice on writing to the gathered crowd. A recurring point in that evening was simple: first drafts are terrible but allow us to experience the "magic of messes."

Lamott's speech occurred in the final days of the NBA regular season. The Golden State Warriors were chasing the historic mark of 73 regular season wins. Steph Curry was also setting a record with every three-point shot. Curry was smashing the previous record of 286 that he set in the previous year. He'd surpassed the mark of 300 made with three-point shots in March, well over one month before the end of the regular season. He'd eventually shoot just over 400.

Lamott's characterization of the writing process and Curry's shooting prowess underscored how accomplishments can be interwoven with missed attempts. Michael Jordan put it this way:

> "I've missed more than 9000 shots in my career. I've lost almost 300 games. 26 times, I've been trusted to take the game winning shot and missed. I've failed over and over and over again in my life. And that is why I succeed."

Anne Lamott's books received acclaim, not the drafts. Steph Curry misses from behind the three-point line over half the time.

How does this apply to data science? There are generally no awards or winning and losing. There are, however, attempts, not at shooting or writing but at insight. To write a book, you must sit and write, even if the result is initially poorly conceived. Steph Curry must be willing to shoot the ball and inevitably miss over half the time in order to score. To gain insight from data, you must form a question, gather data, and perform analysis without a guarantee of helpful results. Lamott is a gifted writer. Curry is an MVP player. Those performing your data analysis should be trained and serious in their exploration. Still, missed attempts can lead to success.

The best written chapters in a draft may not make the final book. Open shots from behind the three-point line may not lead to a score. In mathematics, you may have a clear vision of how to approach a problem. After tackling the question, you may find that it is harder than expected — possibly not even answerable in its current form. Such moments clarify what you know and don't know, and what you can and cannot do.

You may, like when writing a draft or shooting on a practice court, need to step back and do more background work. You may need to learn new ideas or revisit your question. Failing to solve a problem can be disappointing but could also be the step that leads to a bigger insight than expected. In approaching mathematical work, be sure, switching to a baseball analogy, that you are willing to risk a strikeout. As Babe Ruth said, "Every strike brings me closer to the next home run."

In sports analytics, if you perform analysis and don't gain insight, ask another question: why or what can you do differently. Answering this question can enable you to win. Sometimes, we must experience those missed attempts in order to make the game-winning moves that define our success.

10.4. Embracing Imperfection

Using analytics to create successful brackets is an active area of my research. If we consider only one year of data, my work sifts through about 5,000 games involving 350 teams. Analytics enables us to find patterns and trends that might otherwise be overlooked or seem insignificant.

In 2014, Warren Buffett ensured a billion-dollar prize (yes, billion and not million) for anyone who could create a perfect bracket and submit it to the Quicken Loans March Madness online bracket competition. With that many zeros in a prize, I received considerable media attention. In the ESPN online bracket pool, my methods outperformed over 97% of over 4 million brackets in 2009 and over 99.9% of over 5 million in 2010. When the press and public learned about our success, they wanted to know more. Among the various media interviews, I appeared in *The New York Times*, *USA Today*, and on the CBS Evening News.

There is an irony in this attention. The phone kept ringing in the hope of creating that perfect bracket. Yet, on my end, I'd quickly note that even with my work a perfect bracket is unlikely — very unlikely.

If a fair coin toss chooses the winner of every game, then your chances of a perfect bracket are 1 in 9 quintillion. That's 9 with 18 zeros after it. To get a sense of that number, if you could complete 3 billion brackets per second and never repeat what you have already created, it would take almost 100 years to create 9 quintillion brackets.

With such horrible odds, why use analytics? First, my methods do not flip a coin. Historically, people have been about 70% correct in predicting outcomes in the tournament. So, on average, you have a 70% chance of predicting the outcome of a game. In this case, the odds reduce to about 1 in 5.7 billion as opposed to 1 in 9 quintillion. Now comes a key part of using analytics. Suppose my methods give you a 1% improvement in accuracy. Your odds just reduced to 1 in about 2.3 billion. If the accuracy increases to 72%, your odds drop to under 1 in 1 billion. It's still unlikely that my methods will yield a perfect bracket, but the odds alone show the notable improvement.

Data analysis will rarely perfectly describe a phenomenon. Customers will make an uncharacteristic purchase, a stock will fluctuate for an apparently random reason, and people can heal or become ill outside expected parameters.

Turning back to sports, athletics combines luck and skill. Luck, in my mind, is inherently random and cannot be predicted. So, my goal is to increase my ability to quantify skill. Even so, random effects can make my prediction wrong. To see this, think of flipping a coin. If your coin is fair, you have a 50–50 chance of getting heads or tails. However, if you have a weighted coin, you might have a 99% chance of getting heads. If I give you one flip, what do you predict you'll flip? Heads is the best prediction but will not be a perfect one. There is always a chance, since your flips are random, that you may see the outcome of tails.

These dynamics can be true in March Madness. It may be that an underdog has only a 5% chance of winning. Yet, on that day, the odds fall in their favor, and they win. Was my prediction wrong? Yes. Still, in my mind, I made the correct prediction given the underdog's odds of losing.

If I expect perfection in my work, I can race through my analysis without making progress. I may focus only on a perfect algorithm, which, in many applications, is asking me to predict parts of life that are unpredictable. In the process, I may leave strong, interesting results behind due to their incapability to make perfect predictions. It may be that I want a collection of methods for my analysis. Given that each is imperfect, it may be that a battery of approaches helps clarify expectations for the future. Yet, even here, I rarely find that my results create perfect insight.

As you read this section, find out what sporting events will be played over the next week? Whether they are played on the basketball court, a cricket, football, or baseball field, a soccer pitch or in a hockey rink, make your predictions. If you get them all right, keep track of how you made your choices — not because you'll be perfect every week but because you might have found a way to be better, possibly much better, than the average given your analysis. Also, it may be that you got lucky. Embrace being imperfect with your analysis and you might get a better analysis to improve your work.

10.5. Dealing with Enormity

In various parts of this book, we've seen examples of a lot of data existing in our world. Data is an intricate part of today's world. Big data continues to emerge and evolve — so much so that even the concept of what constitutes big data varies. Originally, there were 3 V's of big data — volume, velocity and variety. Soon, there were 4 V's and then 5 and before long 7. Let's start with the original 3 V's of big data.

The volume of big data is easy enough to see. The content of the entire World Wide Web is estimated at upwards of 1 zettabyte, which is 1 trillion gigabytes. That's 100 million times larger than the print collection of the Library of Congress as given in a 2000 study by professors of Berkeley's School of Information Peter Lyman and Hal R. Varian. (Varian would later become Chief Economist at Google.) Now, the Library of Congress is 2,100,00 square feet. And 100 million Library of Congresses is 38 times the surface area of the Earth.

The velocity of data isn't surprising, as we often encounter it. Here are estimates, which may even change by the time of this publication, as to the amount of data happening every day.

- In one minute, over 3.5 billion Google searches are conducted worldwide each minute every day. The average Google search is three words so that's about 100,000 copies of *War and Peace* just in Google searches.
- Every day 293 billion emails are sent. It's been estimated that the average email is around 500 words. But maybe that seems long. So, let's take 200 words as the average length of an email. If we assume each email is 200 words, then approximately 100 million copies of *War and Peace* are sent every day! I measured my copy of *War and Peace* and it came to 2.25 inches. So, every day, if you lined copies of War and Peace side by side,

100 million copies would reach from Denver, Colorado to Honolulu, Hawaii with 200 miles to spare!

- Data is created by many devices today including mobile devices, smart TVs, cars, and airplanes. Collectively, it's estimated that 2.5 exabytes of data are created per day. It's also been estimated that storing every word spoken by humanity over all time would take 5 exabytes.

Finally, there is variety. A decade ago, data fitted neatly into columns and rows but not now. Data comes in multiple, many unstructured forms such as tweets, status updates, wearable device information, videos, and images. When combined with the volume of such data, the landscape of mining for insight has changed dramatically.

This original list grew to 7 V's. In fact, just like our velocity estimates may change by the time you read this, so might the number of V's. The seven, though, are: volume, velocity, variety, variability, veracity, visualization, and value. A quick Internet search can allow you more reading on this expanded list.

I'd like to offer another V — vertigo. It is easy to be overwhelmed by today's data and more importantly, analyzing it. A few years ago, I met with an engineering firm to discuss their robust datasets. A lead member of the group commented, "It's mind boggling to think what all we can mine from our data — so much so, that we don't know where to start."

Rather than ask probing questions about their data, visions of the use of analytics, existing work, or current projects, I asked the same question that I ask my undergraduate student researchers, "What's the simplest thing we can do first? If we can't do that, we'd need to step back before we can do anything."

Have a dataset that might offer valuable insight? Are you stuck deciding where to begin? Look for a topic or question that, if answered, could offer interesting insight but is simple enough that if not answered it would indicate you don't have a good enough handle on your data to answer most other questions.

Data analytics, especially of large datasets, is as much exploration at some stages as analysis. Answering simple questions can give hints as to informed, next steps to take. Wanting to see the end product is natural but waiting for a clear vision before proceeding can be prohibitively limiting. Not knowing what to do with a large or even small dataset isn't unusual.

Such uncertainty can indicate that you are about to embark in a new direction possibly offering innovative insight. Standing in the unknown can make your intellectual world spin. If you do get such a vertigo, I challenge you to focus your thoughts on what will be the next step, however small.

10.6. Get Graphic

The amount of data available for analysis is a modern and dynamic phenomenon. Analyzing data to encourage action is not new. The massive datasets of today make initial analysis mandatory in order to ensure its quality and reliability. This, too, is not new. As mentioned in Chapter 3, famed statistician John Tukey published a highly influential book in 1977 entitled *Exploratory Data Analysis* in which he recommended starting data analysis with graphical tools. In particular, he encouraged constructing and viewing a box and whisker plot to see the maximum, minimum, median, and first and third quartiles. This type of graphics would signal skewed rushing data like Barry Sanders runs discussed in Section 8.1. Often, graphics offer an analyst valuable insight and intuition that directly lead to insightful analytics. Well-crafted visualizations can also tell a story.

As an example, we turn to the Crimean War, which occurred from 1853 to 1856 and led to over half a million casualties on all sides. The mother of modern nursing, Florence Nightengale, volunteered during this war. She compiled data on the deaths of soldiers as she was especially concerned about the unsanitary conditions of the hospitals and their impact on many lives.

To present her work, she developed a circular histogram that she called a "Coxcomb" as seen in Figure 10.1. Nightengale presented her graph to Queen Victoria and to Members of the Parliament and civil servants. The graphics clearly unfolded the story of soldiers' deaths. The lightest gray represents deaths by wounds. The next lightest shade of gray corresponds to deaths by disease and black depicts deaths by other causes. The graph clearly demonstrates that far more soldiers died from preventable diseases than from their battle wounds. The visualization led to improved sanitary conditions in the military hospitals, which resulted in greatly reduced numbers of deaths in the remainder of the war.

Modern graphics are often created to be informative and to allow for a quick understanding of the underlying data. When you use graphical tools

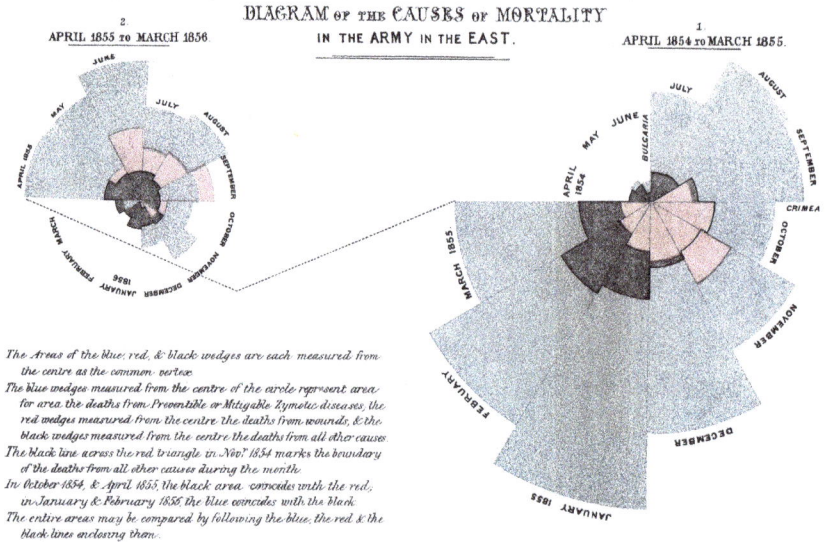

Figure 10.1:   A Florence Nightengale graphics detailing deaths in the Crimean War.

to communicate or analyze data, you rely on how brains process graphical information. The *Power of Infographics* by Mark Smiciklas states that 50% of the brain is dedicated directly or indirectly to visual functions. Further, modern tools allow a user to explore data through graphical tools. For example, my sports analytics group that supplies data for the men's basketball team had a spring meeting with the coaches to review the season and plan for the next year.

Like Nightengale, we prepared a graphical tool for the presentation. One of my student analysts presented an interactive graphics and quickly outlined how to read the visualization. Moments later, an assistant coach asked, "Can we explore the middle region of the graph a bit more? That's interesting that those two players' interactions are so different than the rest." The student moved to another visualization tool and rendered an image utilizing data from the earlier graphics. We all sat silently and then the coaches said, "That would be why we made a rule during the season that such a shot could not be taken." The graphics made it very clear that the team had missed shots of a certain type. The team, in fact, missed them every time. Note that, the coaches didn't receive new insight. But the analysts strongly affirmed a decision they made during halftime in a game and then kept in place for following games. Even if the graphics had preceded the coaches' insight, it would be important to supply a supporting analysis. Due to our meeting, the team planned to further analyze this type of shots and see if they might improve during the off-season.

Do you want to make a point with your analysis? Are you stuck and looking for new insights on data that you strongly believe contains more information to be mined? Consider exploring graphically through various visualizations. You may find that your graphical tools offer new insight on data that might otherwise be difficult, if not impossible, to see.

10.7. A Bit of Humanity in Mathematics

Chapter 9 was very random. We explored methods that make extensive use of random numbers in carefully designed games of chance. At a first glance, it can seem that something random is inherently random and as such unpredictable. When the randomness is designed mathematically, such games of chance can offer insight. In particular, simulation enables us to model human behavior. Keep in mind, however, that humans behave in complex ways and when we make models, we simplify. Analytics, at large, even the data itself, is a model of human behavior.

Data is straightforward. Each bit is a 1 or a 0. Each byte consists of 8 bits. From there we get kilobytes, megabytes, gigabytes, terabytes and the once unfathomable petabytes. Data can encode comments posted on Twitter or the movement of an athlete on a field. The possibilities can seem endless. At the same time, data grows at unprecedented rates. Beyond simply the volume and velocity, data can be frustrating to mine for another reason — it often encodes human behavior.

In my time as a Chief Researcher for the big data company Tresata, I offered advice on their existing projects, which involved a lot of listening. In one such meeting, a data engineer introduced me to a retail project — automating grocery lists. From your shopping data, they figure out how you've shopped in the past and then predict when you'll need items and even what deals would be important to you in the future.

Quickly, I thought of my trips to the grocery store. I shop on one particular day. In fact, my routine is so consistent that the employees at the deli will stop and check the calendar if I come on another day of the week! I buy a handful of products every week and other products periodically. There are a number of items I always stock up on when they are on sale.

I noted these tendencies to the data engineer. He smiled and responded, "You do have a pattern. But, how consistent are you? Do you ever get milk, for instance, somewhere else? Remember, there are only 52 weeks in a year so even some variance can look significant, if we aren't careful." The point was well timed. That very morning, I had zipped down to the local pharmacy to buy milk. The shopping trip secured cereal as a quick breakfast for my family as I left for my early morning commute to the downtown office.

Take a moment and think of your own shopping. How consistent are you? Do even your most consistent purchases have inconsistencies? My grocery list, for instance, changes significantly any time my children are on vacation whether it be Thanksgiving or summer break.

Our world has significant amounts of data and much of it records human experience. While we, as people, have routines and tendencies, we also have the ability, in a moment, to make a different decision based on our perceived context in situations. Given our nuance as humans, our data is nuanced, which has important implications on data analysis.

Just like my trips to the grocery store have small variations, we as people rarely approach situations in only one way. As such, data analysis, even when it gives an average tendency of one person or over a group of people, tells the whole story. As you approach results from data, aggregated results can be helpful. Yet, sometimes, it is the anomaly or slight differences that unravels new insight.

To automate my grocery list, you will need to understand my weekly behavior and also its variances. If I don't buy milk on a given week, it is likely that I had another early pharmacy run. An algorithm that recognizes this understands my shopping and can produce lists that can enrich my trips to the grocery store, possibly reminding me of those items I'll inevitably forget.

When you examine data, be it large or small, look for cases when it comes from human experiences and behavior. Strong tendencies can be important and enlightening. Yet, the nuances are what often uncover the humanity of things. Nuanced analysis often better reflects the human experiences recorded in those 0s and 1s.

These are only some of our lessons on analytics and mathematics. Many of the lessons derive from my writings for the *Journal of Corporate Accounting & Finance*, which is a journal directed to CEOs and corporate accounting and financial executives. We'll learn more in Chapter 19. The lessons of this chapter may give you some ideas for pondering and also springboard you into your own insights from the field. So, let's jump back to our X Games and see what we learn as we train in sports to play in mathematics.

Chapter 11

Thinking in Angles

If we think of a point one unit in the positive direction on the x and y axes, this lands us in the Cartesian coordinate $(1, 1)$. Generally, we don't specify that we have a Cartesian coordinate but in this chapter, we'll switch to polar coordinates, which are another way to express such a location. Polar coordinates have the following relationship with Cartesian coordinates:

$$r = \sqrt{x^2 + y^2} \quad \text{and}$$
$$\theta = \tan^{-1}\left(\frac{y}{x}\right).$$

As seen in Figure 11.1, r denotes the distance of the point from the origin and θ measures the counterclockwise angle of the point from the positive x-axis. So, the Cartesian point $(1, 1)$ is expressed in polar coordinates as $(1, \pi/4)$.

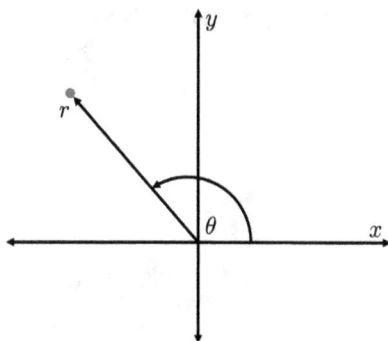

Figure 11.1: Expressing a point (x, y) in polar coordinates (r, θ) is simply an alternative way of locating a point in the plane. Image adapted from [59].

11.1. Polar Position

Let's move to polar coordinates by distorting an image, almost like warped mirrors at a carnival. Every pixel in a photo will have an (x, y) location, which can also be expressed in polar coordinates as (r, θ). We'll then move each pixel in the picture to a new location in which the point at (r, θ) is relocated to (r^2, θ). Let's take the pixels in the image of Muhammad Ali seen in Figure 11.2. We begin by scaling the image so the top right corner is at the point $(1, 1)$ and the lower left is at the point $(-1, -1)$. In this way, every pixel is represented by an ordered pair (x, y) where $-1 \leq x \leq 1$ and $-1 \leq y \leq 1$.

To see how pixels are relocated, let's consider a pixel located at $(x, y) = (0.3, 0.4)$. First, we convert the point to polar coordinates, giving us $r = \sqrt{x^2 + y^2} = \sqrt{0.3^2 + 0.4^2} = 0.5$ and $\theta = \tan^{-1}(y/x) = \tan^{-1}(0.4/0.3) = 0.9273$. In polar coordinates, our point is $(r, \theta) = (0.5, 0.9273)$. We move the point to $(0.5^2, 0.9273) = (0.25, 0.9273)$. So, our point relocates at the same angle but closer to the origin.

Relocating all the pixels in Figure 11.2 results in the image in Figure 11.3(a), appearing almost as if Sonny Liston had punched Ali with a cartoonish effect in their 1964 fight. The 1964 bout didn't need such a punch to be historic. Ali had not yet changed his name from Cassius Clay.

Figure 11.2: Muhammad Ali.

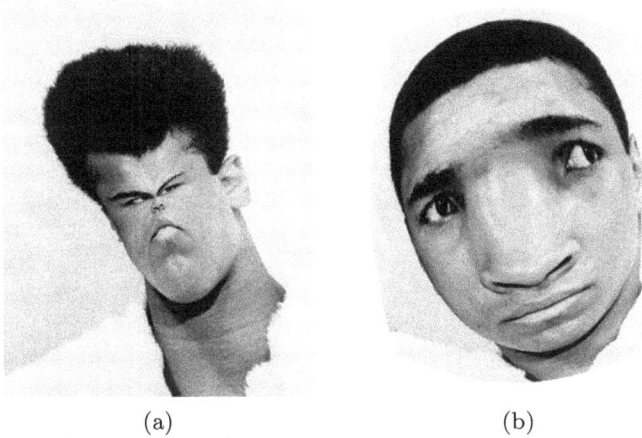

(a)   (b)

Figure 11.3: Warped images of Figure 11.2.

The Liston–Clay fight was named by *Sports Illustrated* as the fourth great-est sporting moment of the 20th century. Clay was a 7–1 underdog and won when Liston gave up at the opening of the seventh round. They met for a second fight just over a year later with Ali winning with a first-round knockout.

Now, let's try another transformation. This time a pixel located at a point (r, θ) will relocate to $(\sqrt{r}, \theta)$. Let's again see where the point $(x, y) = (0.3, 0.4)$ relocates. As we found, this point is $(r, \theta) = (0.5, 0.9273)$ in polar coordinates. Our transformation moves the point to $(\sqrt{r}, \theta) = (\sqrt{0.5}, 0.9273) = (0.7071, 0.9273)$. Yet again, points stay at the same angle but now move farther away from the origin under this transforma-tion. Applying this to the image of Ali in Figure 11.2, we get the image in Figure 11.3(b).

Looking for another effect? Experiment more in polar coordinates to create your own computational carnival mirror.

11.2. Angling an MLB Ballpark

Polar coordinates also enable us to draw curves. For example, a circle of radius 2 centered at the origin is simply the graph of $r = 2$ in polar coordinates. We can also compactly write an equation of a spiral as $r = \theta$. In Figure 11.4(a), we see the curve formed by the equation $r = \theta$.

(a)   (b)   (c)

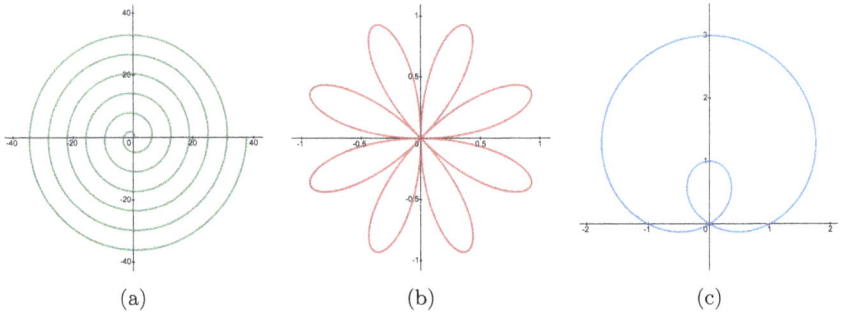

Figure 11.4:   Curves in polar coordinates.

In Figure 11.4(b), we have a rose curve created from $r = \sin(4\theta)$. Finally, Figure 11.4(c) derived from the equation $r = 1 + 2\sin(\theta)$ is called a limacon with an inner loop.

In many ways, polar coordinates are natural in baseball. A hit will exit the bat at an angle from the first base line. If the ball travels a certain distance, it will be a home run. That distance can be thought of as r in polar coordinates. The angle is θ.

Let's use polar coordinates to graph a Major League ballpark. It helps that ballparks have yard measurements to their outfield fence. We'll think of the fence as a sequence of line segments, which you can see in the diagram of Oriole Park at Camden Yards in Figure 11.5(a). Fans have found that the measurements may not exactly match their work. Still, many do. At Camden Yards, the distance to the right field fence down the first base line is, indeed, 308 feet from the home plate. That line extends until it reaches an angle of 25.5 degrees (or 0.445059 radians) from the first base line at a distance of 385 feet from the home plate. Note how we are using angles and distances from the home plate.

How do we express a line segment as an equation in polar coordinates? Let's look at the line $y = mx + b$. We can quickly make our polar substitutions to get $r\sin(\theta) = rm\cos(\theta) + b$. Solving for r gives us $r = b/(\sin(\theta) - m\cos(\theta))$. We have the points in polar coordinates $(318, 0)$ and $(385, 0.445059)$, which are expressed as $(318, 0)$ and

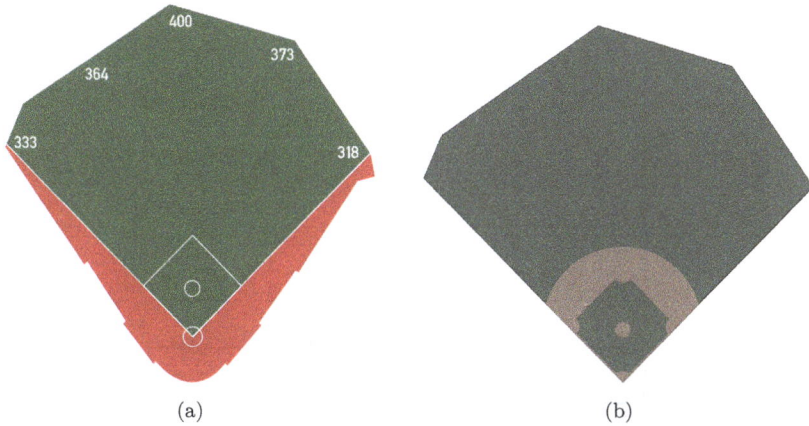

Figure 11.5: Oriole Park at Camden Yards via a diagram of the park (a) and as drawn with polar curves (b).

(347.4, 165.7), respectively. The slope between these points is 5.61942. This gives us the line $y = 5.61942x + b$. Recall that the point $(318, 0)$ lies on the line, so the equation must satisfy $0 = 5.61942(318) + b$ or $b = -1786.98$. So, the outfield fence can be described in polar coordinates as the curve $r = -1786.977/(\sin(\theta) - 5.61942\cos(\theta))$ for $0° \leq \theta \leq 25.5°$. Continuing in this way for the entire outfield fence, gives a curve formed by a sequence of line segments also described by the set of equations:

$$r = \begin{cases} \frac{-1786.977}{\sin(\theta)-5.61942\cos(\theta)}, & \text{if } 0° \leq \theta \leq 25.5° \\ \frac{801.702}{\sin(\theta)+1.830\cos(\theta)}, & \text{if } 25.5° \leq \theta \leq 49° \\ \frac{359.7761}{\sin(\theta)+0.187168\cos(\theta)}, & \text{if } 49° \leq \theta \leq 82° \\ \frac{331}{\sin(\theta)-0.396914\cos(\theta)}, & \text{if } 82° \leq \theta \leq 90°. \end{cases}$$

Plotting this curve along with other parts of the field gives us Figure 11.5(b).

Oriole Park at Camden Yards was chosen since it has only four line segments. Other parks use more. For instance, consider the set of equations that represent the curve for Fenway Park in Boston.

$$
r = \begin{cases}
\dfrac{-119.0423}{\sin(\theta)-0.3941798\cos(\theta)}, & \text{if } 0° \le \theta \le 3.8° \\[2mm]
\dfrac{-402.289}{\sin(\theta)-1.17404\cos(\theta)}, & \text{if } 3.8° \le \theta \le 4.5° \\[2mm]
\dfrac{-808.953}{\sin(\theta)-2.274195\cos(\theta)}, & \text{if } 4.5° \le \theta \le 6° \\[2mm]
\dfrac{-2332.79083}{\sin(\theta)-6.3601456\cos(\theta)}, & \text{if } 6° \le \theta \le 7.1° \\[2mm]
\dfrac{-20759.85313}{\sin(\theta)-55.616\cos(\theta)}, & \text{if } 7.1° \le \theta \le 8.1° \\[2mm]
\dfrac{1129.33168}{\sin(\theta)+2.875435\cos(\theta)}, & \text{if } 8.1° \le \theta \le 31° \\[2mm]
\dfrac{-417.143116}{\sin(\theta)-1.8849057\cos(\theta)}, & \text{if } 31° \le \theta \le 33.8° \\[2mm]
\dfrac{431.2604}{\sin(\theta)+0.587157\cos(\theta)}, & \text{if } 33.8° \le \theta \le 52.2° \\[2mm]
\dfrac{2077.8716}{\sin(\theta)+7.751356\cos(\theta)}, & \text{if } 52.2° \le \theta \le 53.1° \\[2mm]
\dfrac{306}{\sin(\theta)+0.00577087\cos(\theta)}, & \text{if } 53.1° \le \theta \le 90°.
\end{cases}
$$

This leads to the field graphed in Figure 11.6. Note that the left field fence stands at an imposing 37 feet and 2 inches. It's known as the Green Monster and can be seen in Figure 11.6(b). Look carefully at the middle of the picture and you'll see a fan sitting in the set of seats at the top of the fence. To graph baseball fields, this section has used the measurements established by Andrew Fox in his entry "Complete Outfield Dimensions" as posted on FanGraphs in 2015 [31]. To create his measurements, Fox used a variety of online park diagrams. Some fans have used Google maps to establish the distances.

11.3. Global Projection

In this book, projecting a sphere can bring about images of a baseball being hurled to home plate or a soccer ball zipping toward the goal. In this section, we will project a sphere for image manipulation. We begin with a panoramic image, as seen in Figure 11.7(a). In this section, we see how you can take the panoramic image and produce the image in Figure 11.7(b). In particular, we think of the image in Figure 11.7(a) as existing on a sphere and project that image onto a plane that forms the image in Figure 11.7(b), which is called stereographic projection.

(a) (b)

Figure 11.6: Fenway Park in Boston as graphed with polar coordinates (a) and looking at the left field fence known as the Green Monster (b). Green Monster image by Aidan Siegel.

(a) (b)

Figure 11.7: Stereographic projection can turn an image like (a) into what is often called a tiny planet image as seen in (b).

Stereographic projections are common in cartography, which you may have seen in maps of the Earth or celestial charts. The advantage of this type of map is that it preserves angles but distorts areas, which is helpful for navigation, which often involves angles.

We'll create our projection from a pole directly above the plane on which the final image will be created (Figure 11.8). Given this, the upper hemisphere, which correlates to the upper portion of the image will have the most distortion. We can reference any point on the sphere's surface with two angles representing the latitude ϕ and longitude λ, where:

$$-\pi < \lambda < \pi, \quad -\pi/22 < \phi < \phi/2.$$

We'll let x denote the width and y the height of the image and further make our sphere have radius 1 and place the image such that $-1 < x < 1$, $-1 < y < 1$. Since the horizontal units are the longitudes (0–360 degrees) while the vertical units are the latitudes (-90 to $+90$ degrees), we begin with an image that is twice as wide as it is tall.

We will create a ray to calculate the projection of a single point from the sphere to the plane. This ray begins at the projective origin, passes through the sphere, and points at any other point on the surface of the sphere. This ray

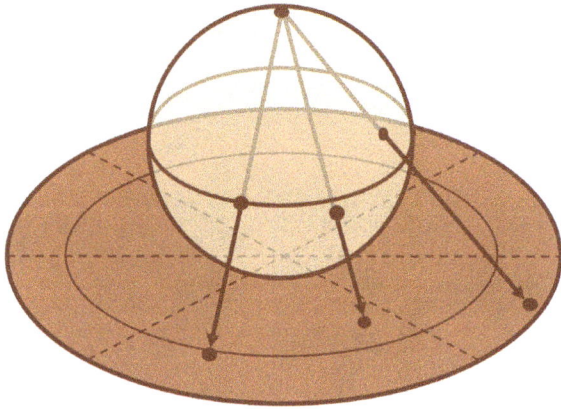

Figure 11.8: Projecting a sphere onto a plane.

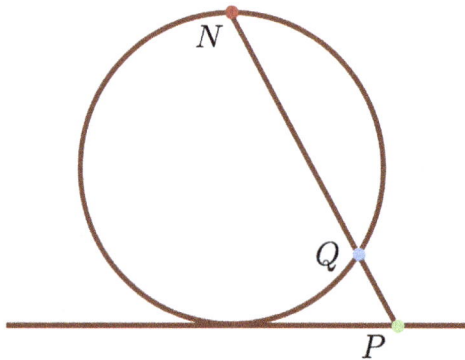

Figure 11.9: Projecting a circle onto a plane.

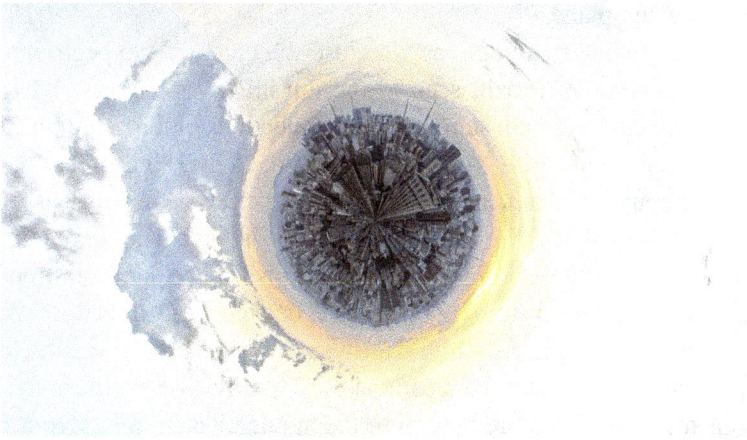

Figure 11.10: Panoramic of Sao Paulo, Brazil from https://www.goodfreephotos.com.

In Figure 11.9, we see the projection in two dimensions. This can help create the necessary formulas. Let λ_0 be the central longitude and ϕ_1 the central latitude. Then, our projected coordinates can be stated mathematically as

$$u = k \cos \phi \sin(\lambda - \lambda_0),$$

$$v = k(\cos \phi_1 \sin \phi - \sin \phi_1 \cos \phi \cos(\lambda - \lambda_0)), \text{ and}$$

$$k = \frac{2R}{1 + \sin \phi_1 \sin \phi + \cos \phi_1 \cos \phi \cos(\lambda - \lambda_0)},$$

where the sphere has radius R and k determines the distance from the sphere to the projected plane.

Let's see another example as we close the chapter. We begin with an image of Sao Paulo, Brazil, seen in the top image of Figure 11.10. The two images below are both stereographic projections of the panoramic. The bottom image is created by rotating the top image by 180 degrees.

Throughout this chapter, we've worked in 2D. An inherent aspect of Cartesian coordinates is the requirement that positive and negative numbers are needed to describe all 2D points. How far is Cambridge from London? It's about 50 miles north. We just gave a distance and direction. Where did you grow up? Few of us would answer with the latitude and the longitude. When do you use polar coordinates? Usually, you will turn to polar coordinates when it's more natural to the application or makes your analysis much easier. Changing your mathematical perspective can open your options.

Chapter 12

Calculus — Just Do It!

If a greatest hits compilation of calculus were written, it would have two main chapters — slope and area. Let's discuss slope first. When we find slope, we can find velocity. When you are driving a car, you can find your current speed by checking the speedometer as seen in Figure 12.1(a). Note that the speedometer is in kilometers/hour, in case you were dreaming of racecars and the German autobahns where the sign in Figure 12.1(b) indicates the end of all restrictions (including speed limits). This is an important lesson of calculus and mathematics; in general, be sure that you know your units of measurement.

12.1. Calculus of Usain Bolt

Calculus helps us when, instead of a speedometer, we have a function giving the position of an object at any time t. In the 2008 Olympics in Beijing,

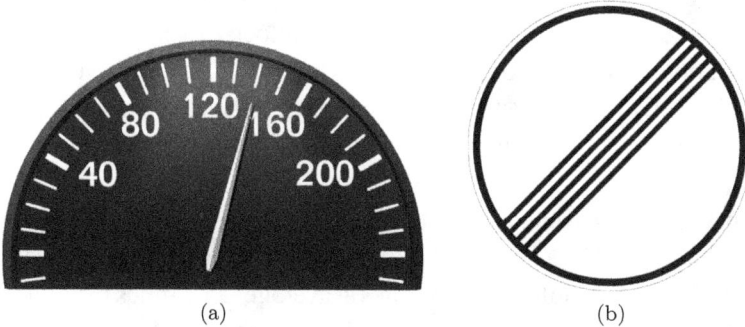

(a)   (b)

Figure 12.1:   Speedometer and a German traffic sign indicating no speed limit.

Table 12.1: Time and distances, excluding the 0.17 second reaction time to the gun, for Usain's world record setting run in the 2008 Olympics.

Time	0	1.68	2.70	3.61	4.48	5.33	6.15	6.97	7.79	8.62	9.52
Distance	0	10.0	20.0	30.0	40.0	50.0	60.0	70.0	80.0	90.0	1.0

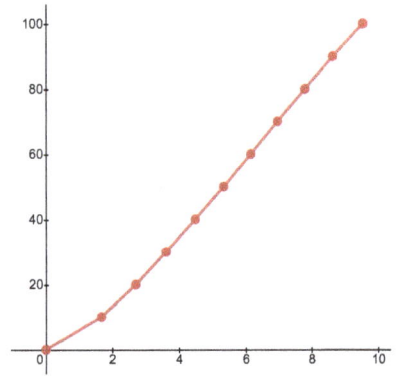

(a)   (b)

Figure 12.2: (a) Usain Bolt in his classic pose, image by Fernando Frazžo/Agência Brasil; and (b) a graph of his split times for his record-setting 100 m run in the 2008 Olympics.

Usain Bolt ran 100 meters in 9.69 seconds, establishing a new world record. Sebastian Schreiber, Wayne Getz, and Karl Smith in their textbook *Calculus for the Life Sciences* give split times, as seen in Table 12.1, during the historic race [54]. If we plot these points and connect them with line segments, we get the graph in Figure 12.2 (b). Calculus helps establish that the slope of each line segment is Usain Bolt's average velocity over that time period. For example, between 3.61 and 4.48 seconds, Bolt's average velocity was $(40 - 30)/(4.48 - 3.61) \approx 11.49$ meters/second. If we convert this to miles per hour, we see that Bolt was running about 25.7 miles per hour at that point in the race.

But computing average velocity isn't quite like looking at a speedometer. A speedometer tells us our speed at that moment. We are finding the average velocity. Although Bolt may have run, on average, 25.7 miles per hour from 3.61 to 4.48 seconds, he didn't necessarily run at that speed the entire time. This is where calculus comes in. We want the velocity in one instant, which

Table 12.2: Average velocities of Usain Bolt with shrinking intervals of time.

Time A	Position	Average velocity between Time A and 4 seconds
3.9000	33.2804	11.4890
3.9500	33.8543	11.5010
3.9750	34.1416	11.5070
3.9875	34.2854	11.5090
3.9937	34.3574	11.5110
3.9969	34.3933	11.5120

is called instantaneous velocity. For this, we need a function, rather than a finite number of points, that describes where Bolt was at each moment in time. Using the split times, Schreiber, Getz, and Smith derived the equation

$$y = -0.0016942x^5 + 0.046520x^4 - 0.52044x^3 + 3.0905x^2 + 2.0288x.$$

for Bolt's position (y) at any given time (x). How fast was Bolt running at 4 seconds? For this, we can still find the average velocity. But now, we take smaller and smaller intervals of time. Let's start by finding the average velocity from 3.9 to 4 seconds. First, Bolt's position at 4 seconds, according to our formula, was $y(4) = 34.4293$. His position at 3.9 seconds was $y(3.9) = 33.2804$. So, his average velocity over this time was 11.489 meters per second. If we find his average velocity over an interval of time that's half as long, namely between 3.95 and 4, we find an average velocity of 11.5010. Continuing this process produces the table in Table 12.2. Note how the average velocity is getting closer and closer to 11.51 meters per second, which is about 25.7 miles per hour. Calculus also teaches ways to compute his velocity even more directly from the function $y(x)$ by finding the slope of the tangent line of the curve $y(x)$ at $x = 4$.

12.2. Steeler Integration — Just Do It

Let's use a sports logo to tackle the second major pillar of Calculus, which is area. We'll work with the Pittsburgh Steelers logo, as seen in Figure 12.3(a). The shape of the red, blue, and yellow emblems is often called an astroid or, if you want to sound quite mathematical, a 4-cusped hypocycloid. The

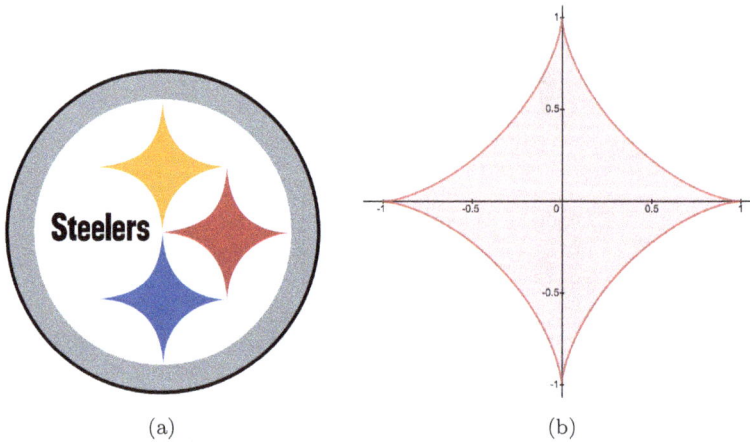

(a)   (b)

Figure 12.3: Watching a game, you can think about the mathematics of the play or of the team logos.

entire curve can be formed parametrically by graphing

$$x = \frac{1}{4}(3\cos(t) + \cos(3t)), \ y = \frac{1}{4}(3\sin(t) - \sin(3t)), 0 \le t \le 2\pi.$$

This produces the curve seen in Figure 12.3(b). The upper right quarter of the asteroid is represented by the function $f(x) = \left(1 - x^{2/3}\right)^{3/2}$ for x from 0 to 1.

Let's find the area of an asteroid by finding the area of $f(x)$, as seen in Figure 12.4(a), which we can then multiply by 4. To begin with, we'll approach this in a similar manner to the way we created mosaics in Chapter 8. First, we draw the vertical lines $x = 0$, $x = 1$, and $y = 0$ since these are the boundaries of our curve. The last line is the horizontal line $y = 1$ since $y \ge f(x)$ for $0 \le x \le 1$. This is our boundary region, which, in this case, happens to be a square.

Now, we'll partition the region into rectangular blocks. In this case, we can form squares since our boundary region is a square. Let's begin by forming five horizontal squares. So, the width of the each square is 1/5 as seen in Figure 12.4(a). In general, if you want n blocks, the width of each block for a left boundary $x = b$ and a right boundary $x = a$ is $(b - a)/n$. For this problem, $b = 1$, $a = 0$, and $n = 5$. If we take $n = 10$, then we get the image in Figure 12.4 (c).

Next, we decide whether to fill a block or not. We'll fill a block if the upper right-hand corner of the block is vertically under the curve. Taking

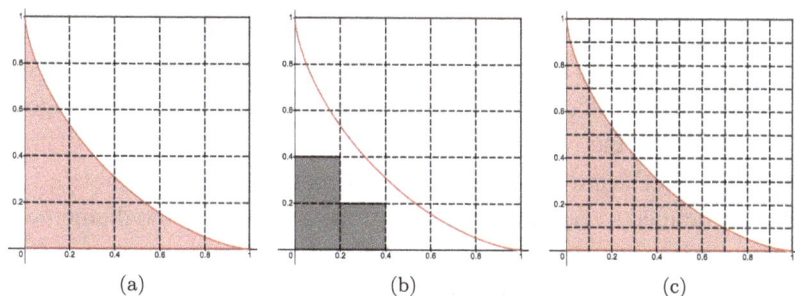

Figure 12.4: Finding the area under a curve is part of Calculus.

$n = 5$ creates the image in Figure 12.4(b). This will give us an approximation for the area under the curve. The entire boundary region has a height and
width of 1. So, it's area is 1. We've filled three of the 25 squares. So, our esti-
mation for the area under the curve is (area of the boundary region)(fraction
of total blocks that are filled) $= (1)(3/25) = 3/25 = 0.12$. We can see
various parts of the red region within blocks that aren't filled. So, not surprisingly, our estimate of 0.12 is too low. However, note if that we increase
the size of n like $n = 10$ in Figure 12.4(c), we get a better estimate of
$21/100 = 0.21$, which we will soon see is closer to the true value. Note that
your estimate might differ since a few blocks render it difficult to entirely
determine, without precise calculation, whether the upper right-hand corner
is vertically under the curve.

This captures the essence of how calculus approaches finding the areas.
Shrink the size of things. Remember how we kept reducing the size of
intervals to find instantaneous velocity. Now, we keep decreasing the
size of the blocks within the boundary region. As we do, we'll find that
our approximated area gets closer and closer to the true value, which is
$\frac{3\pi}{32} \approx 0.29452$.

So,

$$\text{area under the curve } f(x) \text{ and above the } x\text{-axis} = \frac{3\pi}{32}.$$

Calculus uses the following notation to mean the same thing:

$$\int_0^1 f(x)\, dx = \frac{3\pi}{32}.$$

This means "the integral of $f(x)$ from 0 to 1 is $\frac{3\pi}{32}$". You can find the exact value by inputting the integral into some calculators or WolframAlpha.

Let's do more calculus on Steeler's logo. The outer edge of the logo is a circle, which has the well-known ratio of its circumference to its diameter equaling π. Let's find the ratio of the perimeter of the red emblem to its diameter, defined to equal the distance from one point to the diametrically opposite point. The diameter of this asteroid is 2 since two diametrically opposite points are $(1, 0)$, and $(-1, 0)$. We'll find the length of the curve in Figure 12.3 (b) and multiply the result by 4.

The length of the curve $y = f(x)$ from 0 to 1 is

$$L = \int_0^1 \sqrt{1 + [f'(x)]^2}\, dx.$$

Let's see more of the steps and more of the calculus in action. For $f(x) = \left(1 - x^{2/3}\right)^{3/2}$, $f'(x) = -\sqrt{1 - x^{2/3}}/x^{1/3}$. We seem to have a computational pile-up when we consider finding

$$L = \int_0^1 \sqrt{1 + [f'(x)]^2}\, dx.$$

A push of algebra clears a path to the equivalent expression

$$L = \int_0^1 1/x^{1/3}\, dx.$$

This is a much easier expression giving us the result of 3/2. So, the length of a quarter of the astroid equals 3/2, making the perimeter of the complete asteroid 6. We have our desired ratio of the perimeter to the diameter of the asteroid, which equals 6/2 or 3. Things went from being quite irrational with the circle to rational with the asteroids.

Let's take the Nike approach to integration and just do it — or more appropriately, just do more of it by finding the area between two curves. In particular, we will work with the area bounded by $f(x) = 0.6(x - 1)^2 + 1$ and $g(x) = (x - 1.4)^2$. If you solve $f(x) = g(x)$, you find that the region is bounded by $0.239318 \leq x \leq 3.76068$. We see the region in Figure 12.5. Note that we rotated the swoosh so the curves satisfy the vertical line test, allowing us to find functions to define the curves.

Calculus computes such integrals in a very natural way. Note that $f(x) \geq g(x)$ over the entire region. So, the integral of the swoosh would be the

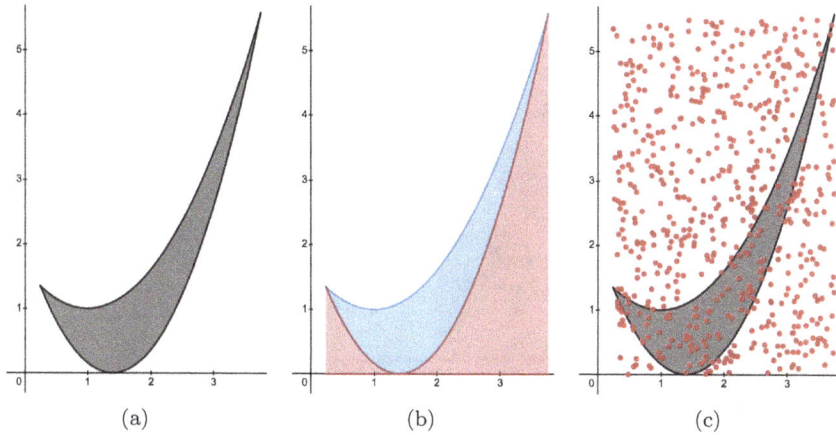

Figure 12.5: Two parabolas can approximate the Nike swoosh.

difference of the integral of f and the integral of g. Graphically, we take the region under the curve for $f(x)$ and above the x-axis, which is the sum of the regions shaded red and blue in Figure 12.5(b). From that sum, we subtract the region under the curve of $g(x)$ but above the x-axis, which is the region shaded red in Figure 12.5(b). This leaves us only with the region shaded blue. Using the notation of calculus, we find that

$$\int_{0.239318}^{3.76068} (0.6(x-1)^2 + 1) - (x - 1.4)^2 \, dx = 2.91099,$$

which WolframAlpha or other computational devices can easily integrate.

12.3. Simulated Calculus

Let's see another way to find the area of the Nike logo in Figure 12.5(a). The logo is the area bounded by $f(x) = 0.6(x-1)^2 + 1$ and $g(x) = (x - 1.4)^2$. This resides in the rectangle bounded by $0.239318 \le x \le 3.76068$ and $0 \le y \le 5.5728$. So, that area of the bounding rectangle is $5.5728(3.76068 - 0.239318) = 19.6238$. To estimate the area of the shaded region, we place random points in the bounding box. Suppose we randomly place T points. Then, we determine how many of those points fall within the shaded region; we'll call this N. Then, our estimated area is $19.6238(N/T)$. For the 600 random points placed in Figure 12.5(c), we get an estimate of 2.9. Want more decimal places? Place more points!

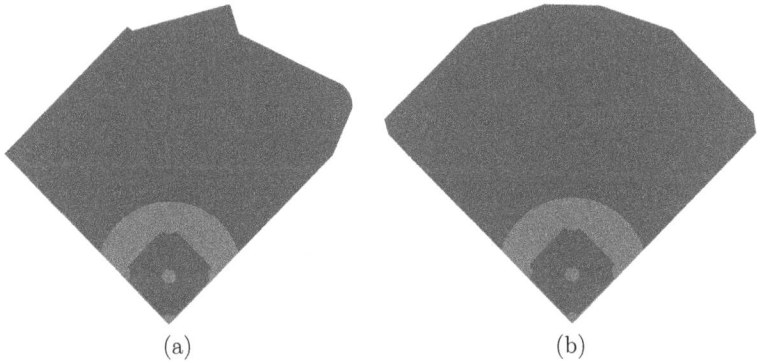

(a)   (b)

Figure 12.6:   Major League Baseball fields vary in their outfield layouts and dimensions.

We are using simulation, like we used in Chapter 9, to integrate. Simulation can help integrate shapes that aren't as quick to integrate by hand as the Nike swoosh, since it was a region bounded by two parabolas. Let's look at Figure 12.6(a); in Section 11.2, we discussed how to produce this image using curves expressed in polar coordinates. This is the Red Sox's Fenway Park in Boston. It's been the home of the Boston Red Sox since 1912. Over its more than 100-year history, there have been renovations and expansions leading to its quirky layout. A home run can be hit by hitting a ball 310 feet from the home plate, which is shorter than other fields. So, the Green Monster was erected, which is a 37.2 foot left field wall towering over the field. Right field has its novelties. There is a lone red seat in section 42, row 37, seat 21. It signifies the longest home run ever hit at Fenway by Ted Williams on June 9, 1946. It was officially measured at 502 feet. The outfield with its layout and quirks is the smallest outfield in MLB.

Fenway hosted the Red Sox when they won the World Series in 1912, 1915, 1916, and 1918. The team would then go 86 years without a championship. Some say, partially in jest, this was due to the Curse of the Bambino. The curse was cast when the team sold Babe Ruth, sometimes nicknamed "The Bambino", to the New York Yankees during the 1919–1920 off-season. Before the trade, the Red Sox were the most successful professional baseball franchise. After the trade, the team would go decades and almost a century without winning a World Series. To add to the curse, the Yankees would rise as one of the most successful sports franchises.

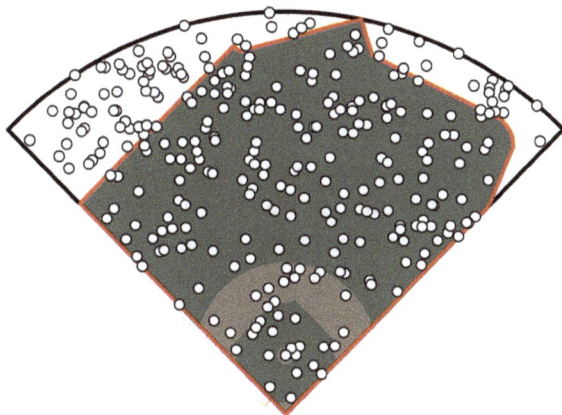

Figure 12.7: To find the area in fair play in Fenway Park, we can use simulation.

In 2004, Boston played St. Louis in the World Series. Boston swept the Cardinals to snap the curse and make sports history. The win was especially notable as the team came back from a 0–3 defeat to the Yankees to win the 2004 American League Championship Series. Busch stadium, as we see in Figure 12.6(b), has a very different layout. It's bigger. How much bigger? How much area did the outfielders have to cover in that series? We will answer this question using simulation.

As with the Nike swoosh, we drop points and see the proportion that falls in the region of interest. For this question, the area of interest is the fair play area of Fenway Park. This is outlined in red in Figure 12.7. The total area where points can fall is contained within the red boundary as well as the black arc seen in the figure. This gives us a quarter circle where points can fall. The points must be dropped with equal likelihood anywhere in the region. When they are, we can take the ratio, as we did with the Nike swoosh. For Fenway, we get an estimate of 2.32 acres when we take enough points to get the same result on multiple simulations.

After finding the area of Busch Stadium, we find that the difference between the two stadiums in terms of total area of fair play is close to 20% of a football field. The Colorado Rockies play in Coors Field, which has the largest region of fair play as given by the Business Insider. The difference between Fenway and Coors fields is almost 30% of a football field. There can be a lot of additional area to cover for outfielders depending on the stadium.

So, we've looked at slope and area. Both approaches contained a fundamental idea that led to the discovery of calculus — divide and conquer. Finding slope and area, in this chapter, we approximated by computing a value over an interval. Then, the interval was shrunk in size. If you study more about Calculus, you'll find that the methods that enable us to quickly find the exact values derive from shrinking the size infinitely many times. Indeed, calculus took divide and conquer to infinity and, in some ways, beyond.

12.4. Free Fall in the Movies

Calculus has become a standard part of the college and advanced high school curriculums. Little did Newton and Leibniz, who originally conceptualized Calculus, know how their discoveries would consume so much pencil led, stacks of paper, and unknown amounts of brain sweat.

An early topic of Calculus is linear free fall motion where an object falls directly down. If an object is dropped from rest, then after t seconds it has fallen $y(t) = -4.9t^2$ meters and has a velocity of $v(t) = -9.8t$.

Let's use this formula in some scenes from movies. First, let's consider the 2017 film *Thor: Ragnarok*. During the film, Dr. Strange opens a portal through which Loki falls after which Dr. Strange and Thor interact. At the end of their conversation, Dr. Strange opens another portal through which Loki falls. After hitting the ground with a huge thud, Loki exclaims, "I've been falling for 30 minutes."

Our formula can indicate how far Loki fell. If we assume his statement gives us an exact timing, then he fell for 30 minutes, which is 1800 seconds. So, he fell $9.8(1800)^2$ meters. If we compute this and convert it to miles, Loki would have fallen 19,730 miles, which is almost two and a half times the diameter of the Earth. That's a hefty fall! That's hard to imagine. To get a sense of how fast he would be falling, let's compute his velocity after 30 minutes. In particular, Loki would be falling $v(1800)$ or 10.9 miles per second when he hit the ground. That's truly superhuman strength! (Figure 12.8).

Keep in mind that this model assumes no air resistance but you never really know where Loki went, so it may not be any more probable that there

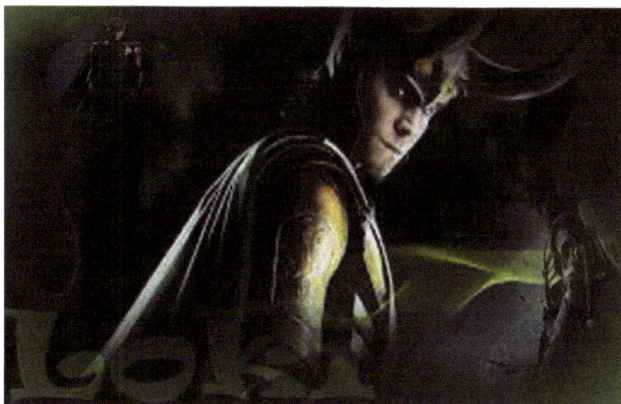

Figure 12.8: Loki takes a hefty fall through a Dr. Strange portal.
Source: https://www.deviantart.com/darkprinceofpersia1/art/Loki-thor-avengers-32826 9175.

is or isn't air resistance. That assumption also means that Loki continually sped up as he fell.

The implications of this model's assumptions have different consequences when we apply our formula for free fall to a scene from the 2002 movie *Lord of the Rings: The Two Towers*. At the very beginning of the film, Gandalf stands at the end of a stone bridge over a deep abyss as he confronts the fiery creature Balrog as seen in Figure 12.9. The wizard's staff pounds against the bridge, weakening the stone structure as it begins to crumble beneath the monster's feet. Balrog descends into the darkness only to extend its fiery whip, which catches Gandalf pulling him into the deep crevasse. If you sit with a stopwatch during this scene (which is of course a natural thing to do), you will find that Balrog's fall lasts about 106 seconds while Gandalf falls about 70 seconds. They meet 17 seconds into Gandalf's descent and end their plunge splashing into a huge body of water deep within Middle Earth. We ask a simple question: How far did Balrog and Gandalf fall?

Let's use our model from Calculus. The only force acting on the two is gravity, which we will simply take to be the same as our Earth. We'll consider changing this assumption in a moment. Note, if Middle Earth, as a celestial body doesn't have gravity, then something quite different is occurring in the physics of that universe! So, we'll consider this a safe assumption. To find the distance Gandalf traveled, we use our formula to

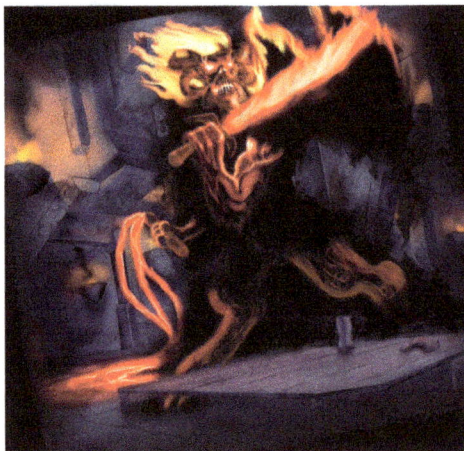

Figure 12.9: In *Lord of the Rings*, Gandalf faces the fiery creature Balrog in Middle Earth.
Source: https://upload.wikimedia.org/wikipedia/commons/1/14/Durin%27s_Bane.jpg.

find a distance of $4.9(70)^2 = 24,010$ meters or 14.9 miles. Repeating this process for Balrog, we compute a total distance of just over 55,056 meters or about 34 miles. That's one huge drop!

Let's reflect on that result. Gandalf and Balrog start their falls at the same spot, landed at the same spot, and fell straight down. Yet, our result states that they fell different distances! Could our assumption on the gravitational pull of Middle Earth be influencing this result? If we denote the constant of gravity by g, then Gandalf fell $(70)^2 g/2 = 2450g/$ meters while Balrog fell $(106)^2 g/2 = 5618g/2$ meters. Even if $g = 0.98$, which is only a tenth the magnitude of our planet's pull, then the difference in the length of their falls is still almost 2 miles!

Keep in mind that our model does not account for the resistance of air. Whatever the density of the air in Middle Earth, it would affect their descents. This makes sense, if we reflect on it. In the absence of air resistance, Gandalf could not have caught up to the creature during their fall. Sky-divers take advantage of air resistance as they change their shape to become more and less aerodynamic to speed up and slow down, respectively.

12.5. Predicting a Soccer Ball's Flight

If we remove the assumption of no air resistance, the model gets more realistic. But how real could such a model get? To answer this, let's look

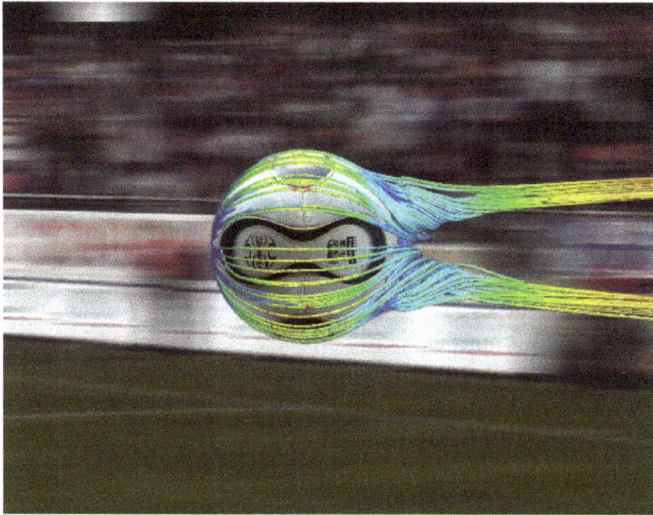

Figure 12.10: The predicted path of a soccer ball.

at soccer, where matches are filled with complex aerodynamics as the ball frequently curves and swerves through the air. The University of Sheffield's Sports Engineering Research Group and Fluent Europe Ltd studied the trajectory of soccer ball, which required a model more complicated than the one we just studied. First, the ball would move in three dimensions rather than simply up and down in the air. The research group focused on shots resulting from free kicks, at which time the ball is placed on the ground and then kicked.

Gandalf and Balrog were calculated to move different distances due to the lack of air resistance. To study a ball moving with air resistance meant the group included what are known as Navier–Stokes equations, which are based on (a) the conservation of mass, (b) conservation of momentum and (c) conservation of energy.

Since 1970, every World Cup has its own official tournament ball. The research group studied four balls with different panel designs, including the 2006 World Cup. The research group found that the shape and surface of a soccer ball, as well as its initial orientation, play a fundamental role in its trajectory.

The research led to predicted paths of kicked balls. Wind tunnel measurements from the University of Tsukuba in Japan verified the results of the studies. The work developed in Sheffield led to detailed analysis of the memorable free kick that resulted in a goal scored by David Beckham of England in a match against Greece during the World Cup Qualifiers in 2001. Beckham's shot left his foot at about 80 mph. The ball cleared the defensive wall of players by about one and a half feet while rising over the height of the goal. As it neared the goal, the ball slowed to 42 mph and dipped into the corner of the net. Calculations showed if the ball had not slowed to this point at that point, the ball would have missed the net and gone over the goal's crossbar.

Analyzing aerodynamics can help a car, bobsled, or bicyclist pass through the air more efficiently. Keep in mind that such models must be developed with care so what we compute mathematically can give us insight on what happens physically.

12.6.   Look like a Calculus Genius

Let's close the chapter with two very quick facts involving integration and differentiation. First,

$$\int_{-1}^{1} x^3 \, dx = 0.$$

You don't even need to know how to integrate. There are two keys to know the answer (which will be zero every time). First, you need what's called an odd function. Let's focus on polynomials. A polynomial is an odd function if every term has odd coefficients. Then, we will only look at integrals where we integrate from $-a$ to a for some value a. In our case, $a = 1$. When this happens, the integral can be instantly stated to equal 0.

Why is the answer 0? Let's look at x^3 from $x = -1$ to $x = 1$ as seen in Figure 12.11. In Calculus, areas above the x-axis are positive and areas below the x-axis are negative. Since x^3 has exactly the same amount of area between -1 and 0 (as seen in red) as between 0 and 1 (in blue) with one region lying above and the other below the x-axis, the total area is 0.

When you integrate an odd function from $x = -a$ to a, the result is always 0. So,

$$\int_{-2}^{2} x^5 - x^3 + x \, dx = 0,$$

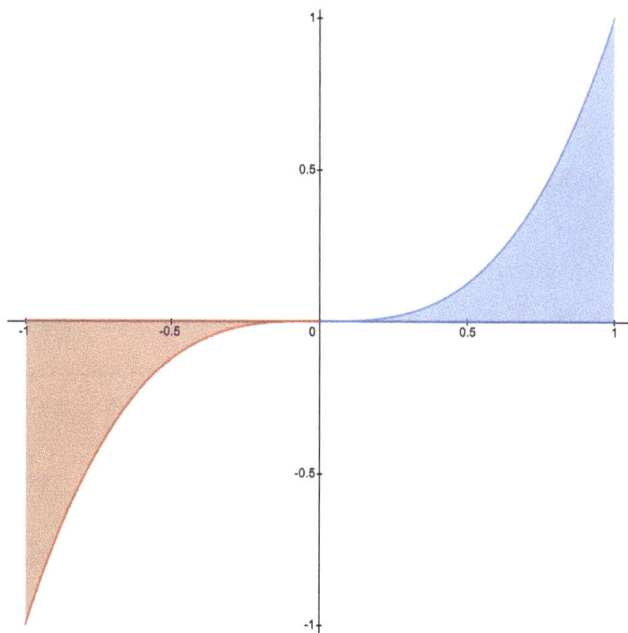

Figure 12.11: The area of an odd function.

since we are integrating from -2 to 2 and the coefficients of each term in the polynomial are odd. Remember that a constant like 5 is not a term of an odd coefficient since $5 = x^0$ and 0 is even. Make it as complicated as you like such as

$$\int_{-101}^{101} 2.11x^{11} - \pi x^5 + \frac{1}{\pi^2} x \, dx = 0.$$

Now, here is a trick related to differentiation. Remember that the slope of the tangent line is the instantaneous velocity of that object. Throw an object up. Differentiation can show that at its peak the object has a velocity (instantaneous velocity) of 0. So, you can say, "I bet I can throw this ball so it will be motionless in the air." Then, you throw it straight up in the air. At its peak, it, indeed, has an instantaneous velocity of 0.

Chapter 13

Passing Secret Notes

Security is an important part of the world. We buy items over the Internet where credit card numbers need to be encrypted. We depend on the privacy of such numbers. This chapter provides insight into some of the issues in encoding information so it cannot be decoded, at least easily. Let's motivate this in sports.

13.1. Cryptic Sidelines

In 2010, the no-huddle offensive evolved in college football. The absence of a team huddle necessitated a new way of sending in calls from coaches who stood along the edge of the playing field. From the sideline came a clever but mysterious signaling system as poster board sized play cards popped up with generally a grid of images like that seen in Figure 13.1. Keep in mind that both teams could see the signs. Among the teams developing the grid of pictures play calling was the University of Oregon, which ran a quick high-tempo offense. Clearly, effective plays were called by the Oregon Ducks with their signs as they marched to the national championship game where they lost by a field goal to Cam Newton and Auburn.

The next season, play cards popped up on the sidelines of many college teams. Fans began paying attention and trying to decipher their meaning. Finding a pattern was tricky as the images came from many sources: emojis, rappers, fast food chain symbols, sports teams, movie characters, and TV celebrities. Figure 13.1 is modeled after a play card used by the University of Kansas during the 2011 season. What play is called with pictures of the Statue of Liberty, the American flag, Abraham Lincoln and the mural

Figure 13.1: College football teams will encode the name of a play with a play card of images.

painted by John Steuart Curry found in the Kansas State Capitol building? Hard to say? That's entirely the point. Could an opposing team figure it out and know the call [37, 58]? To see what's involved in making codes, let's move away from images and work with numbers.

13.2. Caesar's Mathematics

Let's step back in time and share a creative account of a battlefield during the reign of Emperor Julius Caesar. Our story occurs as a battle rages between the armies of Caesar and General Mathematicus Superfluus. Each side is full of brave and valiant soldiers determined to defend homeland and kin.

During the battle, Caesar has immediate need for his courier, Acceleratus. So, he beckons him. Soon, Acceleratus arrives and then a note is placed in his hand. The young courier will follow a perilous route with the intent of taking the precious message, which contains secret information that could ensure Superfluus' defeat, to one of Caesar's generals.

Acceleratus, light of foot and quick in reflexes, darts through the armies. Of Julius Caesar's soldiers, he repeatedly asks for the location of the general who is difficult to locate given the chaos. Meanwhile, General Mathematicus Superfluus' army continues its relentless press on Julius Caesar's army.

Neither army has taken the advantage but suddenly a possibly fateful turn of events occurs. A soldier in General Mathematicus Superfluus' army catches the courier. He returns him to General Mathematicus Superfluus. The note is opened and its contents read:

WKH HQHPB LV ZHDN RQ WKH OHIW IODQQ

Looking deep into the face of the courier, Superfluus senses the importance of the note. He knows well of Caesar's cunningness and senses that the words are somehow code. But how does he determine their contents? He tries threatening the courier. The courier does not divulge the secret. He then decides to entice the courier with a wonderful prize that he might receive as a trade for the true contents of the enclosed code. In time, Superfluus again recalls the wisdom of his foe Julius Caesar. He would never divulge the key to his code to the courier who travels with it. And so, from General Mathematicus Superfluus' army are called some of the finest who are asked to examine the message and decipher its hidden message. They work vigorously as the battle rages on.

Now, let's step into the future where we know the secret code that Caesar used. Taking a message and putting it into a code is called encryption. The steps that exchange letters for encrypted letters are called a cipher. The type Caesar used is also called the shift cipher since each letter is replaced by a letter some fixed number of positions further down the alphabet. To help us use and understand the shift, we'll use the secret decoder ring in Figure 13.2.

Let's begin with a shift of 8. What letter replaces G? Put your finger over the letter G and count 8 letters clockwise around the circle, which will place your finger on letter O. So, the letter G becomes, or is encoded as, the letter O.

What would replace the letter N? What letter would replace the letter X? Try it and then read the answer on the next page. After you have placed your finger on the N and the letter X and counted clockwise around the circle 8 letters, your finger should have landed on the letters V and F. Let's try some more.

Figure 13.2:   Secret decoder ring that creates messages just like Julius Caesar.

Encoding a word is the same process but simply takes a bit longer. Let's use a shift of 5. Then the word

M	A	T	H

becomes

R	F	Y	M

Think about it. You can encrypt messages to friends. Even if they are intercepted, they'll look like gibberish. Remember that your friends should be able to derive the original messages from the encrypted message. This is called decoding. Just let them know how many letters you shifted. This is called the key. It unlocks the secret message.

Decoding is easy. Simply count counter-clockwise around the circle. Let's use a shift of 9 and decode the letter D. Put your finger on the letter D and count 9 letters counter-clockwise. Your finger should land on the letter U. Just like before, we replace letters. The D becomes a U.

Using a shift of 7, then the letters:

K	L	J	V	K	L

Table 13.1: Assigning numeric values to letters to aid in creating codes.

A	B	C	D	E	F	G	H	I	J	K	L	M
0	1	2	3	4	5	6	7	8	9	10	11	12
N	O	P	Q	R	S	T	U	V	W	X	Y	Z
13	14	15	16	17	18	19	20	21	22	23	24	25

decode to be the word:

D	E	C	O	D	E

You can also encode and decode without the circle of letters in Figure 13.2. Each letter is assigned a number where A = 0, B = 1, C = 2, and finally Z = 25 as seen in Table 13.1. Let's again encode MATH with a shift of 5. The letter M is assigned the number 12. We then add the shift to that number, which is $12+5 = 17$ for the letter M. Since R is assigned 17, R becomes our encoded letter. Next, A is assigned 0 and $0 + 5 = 5$, which is assigned the letter F. What if we had the letter Y? This is assigned the numerical value of 24, and $24+5 = 29$. We simply find the remainder on dividing this number by 25, which is 4 in our case. In fact, when we encoded M with an assigned value of 12, $12 + 5 = 17$, which is the remainder of 17 when divided by 25. Mathematically, we say 29 modulo 26 equals 3 and is often written as 29 mod 26. To encode, we find the numeric value x of the letter and encode by computing $(x + k)$ mod 26 where k is the shift. To decode a letter with the numeric value x, compute $(x - k)$ mod 26.

Using the modulus operation, confirm the following when you use a shift of 6.

W	A	X	becomes	C	G	D

Be sure to check that CGD becomes WAX so you practice encoding and decoding.

Suppose General Superfluous remembered that Julius Caesar was known for a fondness of the number 3. Use a shift of 3 to decode the message. Here it is so you don't have to flip back:

WKH HQHPB LV ZHDN RQ WKH OHIW IODQN

and here's some space for your decoded message.

Once you get it, start creating those secret messages! Count carefully. It's hard to decode a shift of 9 if you miscount.

However, note that this isn't all that secure of a cipher. If someone knows this cipher might be an option for your encoding, then they simply need to test all 25 options and see if something pops out that can be read. If you had a long enough piece of text and knew the original text was in English, then you could also find the proportion of times each letter appears. In Figure 13.3, we see the frequency each letter appears in English. The letter "e" is the most common, and appears almost 13% of the time, whereas z appears far less than 1% of time. Note that the Caesar cipher wouldn't change the frequencies of these letters. The cipher simply shifts which letters have the frequencies. With a shift of 1, "f" would appear about 13% of the time. So, Caesar's message was encrypted but, if one had

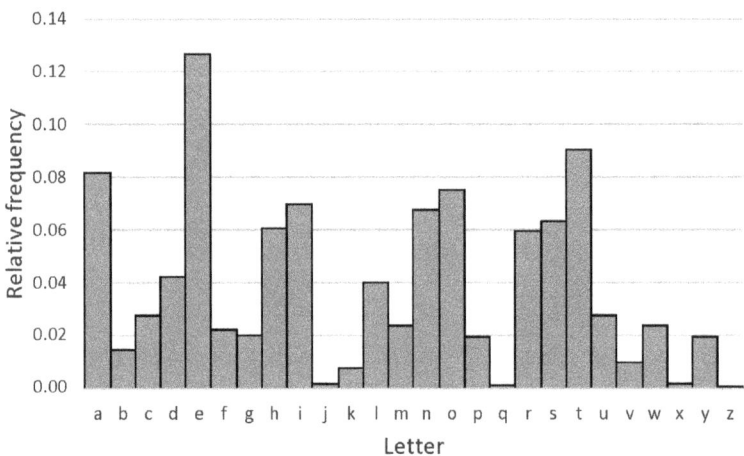

Figure 13.3:   The relative frequency at which letters appear in English.

an inkling about how the cipher worked, could be easily decrypted, even by hand.

13.3. More Advanced Cryptography

Let's learn to send and receive secret messages using another, more secure cryptosystem called the Vigenere cipher. Invented by a Frenchman in the late 16th century, this code was considered unbreakable until the mid-19th century, when Charles Babbage found a way to crack it. You can learn about his method, as well as more modern methods for breaking this cipher, by reading more about cryptography.

Charles Babbage is an important figure in the history of computers. Troubled by errors in published tables of numbers, he designed a machine called the Difference Engine to automate some calculations. He also proposed a more ambitious machine called the Analytical Engine, often considered to be the earliest proposed general-purpose computer. It was never built. (The novel *The Difference Engine*, by Gibson and Sterling, imagines what might have happened in Victorian times had Babbage succeeded.)

The Vigenere cipher uses one table to encrypt a message and another table to decrypt. These tables are shown in Table 13.2. A *key* is also required to perform either operation. A key is a secret word known only to the sender and the intended recipient.

To understand how encryption works, consider the following example. Suppose we wish to encrypt the following message, taken from *The New Hacker's Dictionary* (E. S. Raymond, MIT Press, 1996):

```
program, n. 1. A magic spell cast over a
computer allowing it to turn one's input
into error messages.
```

Suppose the key is `hacker`. To encrypt the first letter in the message, `p`, first compute its position in the alphabet. The letter `a` has position 0, the letter `b` has position 1, which continues giving letter `z` position 25. So, the letter `p` has position 15. Next, compute the position of the first letter in the key, `h`, which is 7. The result of the encryption is the character at row 15 and column 7 of the encryption table, which is `w`.

For the second letter in the message, use the second letter of the key. The letter `r` has position 17 and `a` has position 0, so the result here is the

Table 13.2:   Vigenère encryption (left) and decryption (right) tables.

Encryption table

	0	1	2	3	4	5	6	7	8	9	10	11	12	13	14	15	16	17	18	19	20	21	22	23	24	25
0	a	b	c	d	e	f	g	h	i	j	k	l	m	n	o	p	q	r	s	t	u	v	w	x	y	z
1	b	c	d	e	f	g	h	i	j	k	l	m	n	o	p	q	r	s	t	u	v	w	x	y	z	a
2	c	d	e	f	g	h	i	j	k	l	m	n	o	p	q	r	s	t	u	v	w	x	y	z	a	b
3	d	e	f	g	h	i	j	k	l	m	n	o	p	q	r	s	t	u	v	w	x	y	z	a	b	c
4	e	f	g	h	i	j	k	l	m	n	o	p	q	r	s	t	u	v	w	x	y	z	a	b	c	d
5	f	g	h	i	j	k	l	m	n	o	p	q	r	s	t	u	v	w	x	y	z	a	b	c	d	e
6	g	h	i	j	k	l	m	n	o	p	q	r	s	t	u	v	w	x	y	z	a	b	c	d	e	f
7	h	i	j	k	l	m	n	o	p	q	r	s	t	u	v	w	x	y	z	a	b	c	d	e	f	g
8	i	j	k	l	m	n	o	p	q	r	s	t	u	v	w	x	y	z	a	b	c	d	e	f	g	h
9	j	k	l	m	n	o	p	q	r	s	t	u	v	w	x	y	z	a	b	c	d	e	f	g	h	i
10	k	l	m	n	o	p	q	r	s	t	u	v	w	x	y	z	a	b	c	d	e	f	g	h	i	j
11	l	m	n	o	p	q	r	s	t	u	v	w	x	y	z	a	b	c	d	e	f	g	h	i	j	k
12	m	n	o	p	q	r	s	t	u	v	w	x	y	z	a	b	c	d	e	f	g	h	i	j	k	l
13	n	o	p	q	r	s	t	u	v	w	x	y	z	a	b	c	d	e	f	g	h	i	j	k	l	m
14	o	p	q	r	s	t	u	v	w	x	y	z	a	b	c	d	e	f	g	h	i	j	k	l	m	n
15	p	q	r	s	t	u	v	w	x	y	z	a	b	c	d	e	f	g	h	i	j	k	l	m	n	o
16	q	r	s	t	u	v	w	x	y	z	a	b	c	d	e	f	g	h	i	j	k	l	m	n	o	p
17	r	s	t	u	v	w	x	y	z	a	b	c	d	e	f	g	h	i	j	k	l	m	n	o	p	q
18	s	t	u	v	w	x	y	z	a	b	c	d	e	f	g	h	i	j	k	l	m	n	o	p	q	r
19	t	u	v	w	x	y	z	a	b	c	d	e	f	g	h	i	j	k	l	m	n	o	p	q	r	s
20	u	v	w	x	y	z	a	b	c	d	e	f	g	h	i	j	k	l	m	n	o	p	q	r	s	t
21	v	w	x	y	z	a	b	c	d	e	f	g	h	i	j	k	l	m	n	o	p	q	r	s	t	u
22	w	x	y	z	a	b	c	d	e	f	g	h	i	j	k	l	m	n	o	p	q	r	s	t	u	v
23	x	y	z	a	b	c	d	e	f	g	h	i	j	k	l	m	n	o	p	q	r	s	t	u	v	w
24	y	z	a	b	c	d	e	f	g	h	i	j	k	l	m	n	o	p	q	r	s	t	u	v	w	x
25	z	a	b	c	d	e	f	g	h	i	j	k	l	m	n	o	p	q	r	s	t	u	v	w	x	y

Decryption table

	0	1	2	3	4	5	6	7	8	9	10	11	12	13	14	15	16	17	18	19	20	21	22	23	24	25
0	a	z	y	x	w	v	u	t	s	r	q	p	o	n	m	l	k	j	i	h	g	f	e	d	c	b
1	b	a	z	y	x	w	v	u	t	s	r	q	p	o	n	m	l	k	j	i	h	g	f	e	d	c
2	c	b	a	z	y	x	w	v	u	t	s	r	q	p	o	n	m	l	k	j	i	h	g	f	e	d
3	d	c	b	a	z	y	x	w	v	u	t	s	r	q	p	o	n	m	l	k	j	i	h	g	f	e
4	e	d	c	b	a	z	y	x	w	v	u	t	s	r	q	p	o	n	m	l	k	j	i	h	g	f
5	f	e	d	c	b	a	z	y	x	w	v	u	t	s	r	q	p	o	n	m	l	k	j	i	h	g
6	g	f	e	d	c	b	a	z	y	x	w	v	u	t	s	r	q	p	o	n	m	l	k	j	i	h
7	h	g	f	e	d	c	b	a	z	y	x	w	v	u	t	s	r	q	p	o	n	m	l	k	j	i
8	i	h	g	f	e	d	c	b	a	z	y	x	w	v	u	t	s	r	q	p	o	n	m	l	k	j
9	j	i	h	g	f	e	d	c	b	a	z	y	x	w	v	u	t	s	r	q	p	o	n	m	l	k
10	k	j	i	h	g	f	e	d	c	b	a	z	y	x	w	v	u	t	s	r	q	p	o	n	m	l
11	l	k	j	i	h	g	f	e	d	c	b	a	z	y	x	w	v	u	t	s	r	q	p	o	n	m
12	m	l	k	j	i	h	g	f	e	d	c	b	a	z	y	x	w	v	u	t	s	r	q	p	o	n
13	n	m	l	k	j	i	h	g	f	e	d	c	b	a	z	y	x	w	v	u	t	s	r	q	p	o
14	o	n	m	l	k	j	i	h	g	f	e	d	c	b	a	z	y	x	w	v	u	t	s	r	q	p
15	p	o	n	m	l	k	j	i	h	g	f	e	d	c	b	a	z	y	x	w	v	u	t	s	r	q
16	q	p	o	n	m	l	k	j	i	h	g	f	e	d	c	b	a	z	y	x	w	v	u	t	s	r
17	r	q	p	o	n	m	l	k	j	i	h	g	f	e	d	c	b	a	z	y	x	w	v	u	t	s
18	s	r	q	p	o	n	m	l	k	j	i	h	g	f	e	d	c	b	a	z	y	x	w	v	u	t
19	t	s	r	q	p	o	n	m	l	k	j	i	h	g	f	e	d	c	b	a	z	y	x	w	v	u
20	u	t	s	r	q	p	o	n	m	l	k	j	i	h	g	f	e	d	c	b	a	z	y	x	w	v
21	v	u	t	s	r	q	p	o	n	m	l	k	j	i	h	g	f	e	d	c	b	a	z	y	x	w
22	w	v	u	t	s	r	q	p	o	n	m	l	k	j	i	h	g	f	e	d	c	b	a	z	y	x
23	x	w	v	u	t	s	r	q	p	o	n	m	l	k	j	i	h	g	f	e	d	c	b	a	z	y
24	y	x	w	v	u	t	s	r	q	p	o	n	m	l	k	j	i	h	g	f	e	d	c	b	a	z
25	z	y	x	w	v	u	t	s	r	q	p	o	n	m	l	k	j	i	h	g	f	e	d	c	b	a

entry in row 17 and column 0, which is r. The third letter of the message combines with the third letter of the key to make q, and so on. When we run out of letters in the key, we start over at the beginning. For instance, the seventh letter of the message, m, combines with the *first* letter of the key, h, to yield t.

We'll print punctuation, digits, spaces, and any other characters without any change, and these characters do not affect the current position in the key. For example, after encrypting the first word (program), we print a comma and space, then encrypt the n using the second letter in the key, then print the period, space, number 1, period, and space, then encrypt the next letter (A) using the third letter in the key. We will also turn encrypted text to lowercase, resulting in lowercase output.

The encrypted message is shown below.

```
wrqqvrt, n. 1. c wexpc uzics cccx fcet k
gftpwdii hlnyazug kd xf autx sel's kxtla
ipds vyrqb qvzscqij.
```

Decryption proceeds in exactly the same way, using the second table. For example, the first letter in the message has position 22 in the alphabet, and the character at row 22 and column 7 in the decryption table is p. The decrypted message is exactly the same as the original message, except that the A is printed in lowercase.

You are ready to encode and decode messages using simple arithmetic that has enabled secret messages to be written for literally several thousands of years!

Chapter 14

Measuring the Probable

Probability and percentages play an important role in sports. As we saw in Chapter 9, Monte Carlo simulation can help us to determine probabilities. Since this chapter is on probability, let's use simulation to consider another probability. This time in Major League Baseball.

In 1941, Joe DiMaggio had a 56-game hitting streak. His batting average that year was 0.357. Let's make some assumptions and see the likelihood of this event, at least according to the model we will develop. First, we'll assume DiMaggio was at the plate 4 times in each of the games. Let's assume the probability of DiMaggio not getting a hit was $(1 - 0.357) = 0.643$ each time he was at the plate. So, the probability that he would be hitless four consecutive times at the plate would be $(0.643)^4 = 0.17094$. This helps us to compute the probability of DiMaggio getting at least one hit (the opposite of being hitless), which would be $0.82906 = 1 - 0.17094$ (Figure 14.1).

This is where you get to be Joe DiMaggio and actually have a slightly better batting average. Here's how you play. Pick a number between 1 and 6; let's call it X. To hit like Joe DiMaggio, don't roll X when you roll a six-sided die. Ready? Roll the die 56 times. If you never roll X, then you matched Joe DiMaggio's hitting streak!

Chances are that you didn't do it! The probability of never rolling X in 56 rolls is $(5/6)^{56} = 0.0000368$, which is about once in over 25,000. What if we used DiMaggio's batting average? Then you have about a 1 in 36,000 chance. Quite a feat!

Now, let's try some basketball. Suppose Player A shoots 25% from the free throw line while Player B shoots 75%. Both players get two shots after being fouled. Here's your question.

Figure 14.1:   Joe DiMaggio on a 1939 playing card published by Bowman Gum.

Question: Who is more likely to score exactly one point in their two shots from the free throw line? Player A or Player B?

The answer is a bit too long for a footnote, so it's at the end of the chapter with a necessary spoiler alert. Feel free to keep reading and turn to the end of the chapter when you are ready with the answer.

14.1.   Improbable Perfection

On June 13, 2012, Matt Cain retired all 27 batters he faced, resulting in the 22nd perfect game in Major League Baseball history. As special as it may be for a pitcher to keep all opposing batters from reaching the first base, another, seemingly more improbable, event happened that evening. Behind the plate, Ted Barrett served as the umpire. He wore the black and white stripes and recalled another game pitched by David Cone in 1999, which was, in fact, another perfect game.

A natural response to Ted Barrett's umpiring is: "What are the chances of this umpire calling both games?" Let's estimate through some probability and modeling.

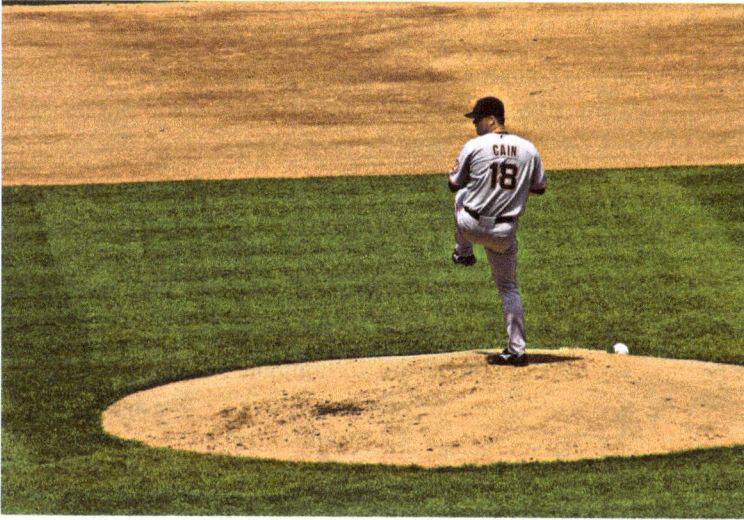

Figure 14.2: Matt Cain, who pitched a perfect game for the Giants in June 2012.

There are approximately 100 umpires in Major League Baseball. Four referees officiate a game. There are 30 MLB teams, so we will assume 15 match-ups occur each "day". As such, there is a 60% chance that an umpire officiates on a given day. Then, there is a 1 in 15 chance that an umpire is selected for a given game. Finally, there is a 1 in 4 chance that an umpire is the home plate umpire, assuming all 100 umpires have an equal chance of serving this role. Since there have been 7 perfect games in the approximately 32,000 games played since the first perfect game pitched by Dave Cone, we estimate the probability that an umpire calls the balls and strikes off a perfect game as

$$(60/100)(1/15)(1/4)(7/32000) \approx 2.1 \times 10^{-6},$$

which corresponds to 1 in 2 million.

Therefore, the probability of serving as the home plate umpire for two such games is

$$(2.1 \times 10^{-6})(2.1 \times 10^{-6}) \approx 4.5 \times 10^{-12},$$

or 1 in 4.5 trillion! How big is this number? Big; 4.5 trillion seconds is over 140,000 years, reaching back to before the last glacial period.

14.2. Yard by Yard in the NFL

In a football game, a team may intercept the ball and begin their possession on the other team's 20-yard line. At that moment, they are in the red zone and with it comes the stats. The announcers will talk about the offense's red zone scoring percentage, which is the percentage in which a team converts their presence in their opponent's red zone into some form of points. What if your team has the ball on the other team's 30-, 22-, or even 21-yard line? Should you be excited? What difference does this make?

To begin our analysis, let's determine the value of where an offensive drive begins. We'll use probabilities to compute the expected number of points a team scores at, for instance, the other team's 20-yard line. Let's see how to do this on a small dataset seen in Table 14.1.

First, we compute the probability of scoring 0, 3, and 7 points, since those are the outcomes. We had four cases where we began on our 20-yard line. Two times we didn't score; once we got a touchdown and once we made a field goal. So, our probability of getting a touchdown, from this data, if you start on the 20-yard line, is 1/4. The probability of not scoring is 1/2 and the probability of a field goal is 1/4. So, if we start on the 20-yard line, the expected number of points is $(0)(1/2)+(3)(1/4)+(7)(1/4) = 10/4 = 2.5$. We only have two instances of starting on the 30-yard line which produced no score and a touchdown. So, the expected points from this yard line is $0(1/2) + 7(1/2) = 7/2 = 3.5$, giving us a sense of how much more likely scoring is from the 30-yard line.

Now, instead of a sample size of six plays, let's use 119, 489 plays of where the ball began for an offensive drive between the 1999 and 2018

Table 14.1:   A small sample of outcomes in football.

Yard line drive begins	Score result
Own 20	7
Own 30	0
Own 20	0
Own 20	3
Own 30	7
Own 20	0

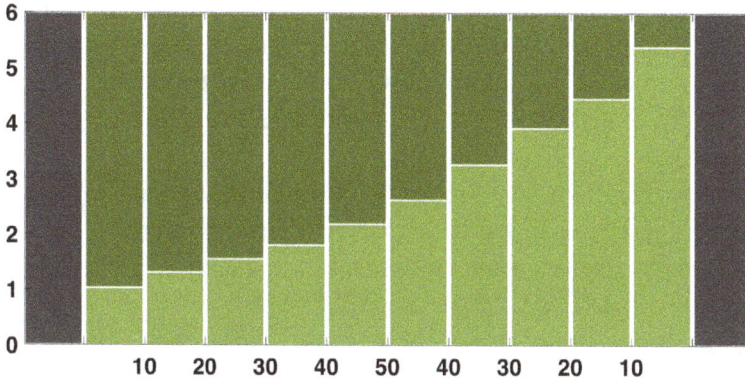

Figure 14.3: Expected points over an NFL football field.

seasons of the NFL in the regular season or playoffs. We see the graph in Figure 14.3. Suppose you begin a drive within your own 10-yard line. This happened 7,991 times and led to a touchdown and field goal 928 and 586 times, respectively. So, a touchdown (which we will take as 7 points) occurred 11.6% of the time and a field goal happened 7.3% of the time. This enables us to find the expected number of points equaling $0.116(7) + .073(3) = 1.03$ expected points. You see this as the first bar chart on the left in our graph.

We see, as expected, that the number of expected points increases as we go from the left to the right. For example, if you start within your 10-yard line, your expected points is 1.03, as we just computed. If your offense gets the ball between the 10 and 20, then you move up to 1.31 expected points. If you get the ball on the other team's 30-yard line, then your expected points are 3.92.

A few other insights are available from this graph. First, when is the number of expected points greater than the value of a field goal? This occurs when you begin an offensive drive within the 40-yard line but beyond the 30-yard line. Your expected points is 3.26. Let's also look at when our expected points equals 5, which indicates that touchdowns are being scored more than field goals and not scoring. For our data, this happens when your drive begins within your opponent's 10-yard line. If you get the ball within the 10, you should be scoring a touchdown, which fans expect as a nonscoring drive would result in loud groans of disappointment in a stadium.

Note that we are only looking at where an offensive drive begins. We could look at the expected points when you are on the 39-yard line with 6 yards to go on second down. You could include the quarter in such an analysis. Each analysis offers insight into the game and helps evaluate outcomes.

Are you ready for some football? You might just want to keep track of where a team places the ball when it begins an offensive drive. This analysis might give you greater insight into who might be raising their arms in victory by night's end.

14.3. To Swing or Not in Mudville's

Baseball is filled with mathematical computations from counting strikes, balls, runs and errors to calculating batting averages, on-base or slugging percentages. Mathematics is an important coaching assistant with these and more advanced analytics such as those documented in the book and film *Moneyball*. The fields of probability and statistics offer well-worn techniques in the math toolkit of baseball. As sportswriter Arthur Daley wrote,

> "A baseball fan has the digestive apparatus of a billy goat. He can, and does, devour any set of statistics with insatiable appetite and then nuzzles hungrily for more."

With so much math ingrained into the game, it is natural to visit the ball field for examples that teach and explore ideas in probability and statistics. We can naturally introduce important ideas such as when to add and when to multiply probabilities, the difference between "at least" and "at most" in a statement of probability, and when events are independent or dependent. An ally through this forest of ideas is the probability tree; it will soon clarify a question posed by a famous tale of the baseball.

The story is set with the home team from Mudville at the plate in the bottom of the ninth with two outs and down by two runs. The seats remain filled since only two hitters stand between that final out and the "Mighty Casey" strutting to the plate. As if the gods of highlight reels were directing the game, the home team from Mudville gets a single and double,

> "Then from 5,000 throats and more there rose a lusty yell; it rumbled through the valley, it rattled in the dell; it knocked upon the mountain and recoiled upon the flat, for Casey, mighty Casey, was advancing to the bat."

So confident in his abilities, Casey doesn't even swing at the first two pitches, which were strikes. And then,

> "They saw his face grow stern and cold, they saw his muscles strain, and they knew that Casey wouldn't let that ball go by again."

Casey shatters the air with his mighty swing and...

> "somewhere men are laughing, and somewhere children shout; but there is no joy in Mudville — mighty Casey has struck out."

First published on June 3, 1888 in the *San Francisco Examiner*, this poem *Casey at the Bat: A Ballad of the Republic Sung in the Year 1888* has echoed through the decades and integrated into the lore of baseball.

The tale of Casey standing at the bat rings true over 100 years after it was penned (Figure 14.4). Fans have sat on the edge of their seats, whether it be at the stadium or at home watching or listening to a game. It is again the bottom of the ninth, two outs, runners in scoring position when the strongest batter on the team steps to the plate. All eyes are focused as the pitcher throws the ball right down the middle of the plate; the batter watches it sail into the catcher's mitt. "Wow", you think, "that was his pitch; that was a missed opportunity." The TV commentary notes how the batter is "feeling

"TEN THOUSAND EYES WERE ON HIM AS HE RUBBED HIS HANDS WITH DIRT"

"BUT THERE IS NO JOY IN MUDVILLE— MIGHTY CASEY HAS STRUCK OUT"

"WHENEVER HOPPER APPEARS BEFORE THE FOOTLIGHTS"

Figure 14.4: The mighty (and arrogant) Casey at the plate in Mudville in an illustration from 1911.

out" the pitcher. Then, a sense of dread sweeps over you as the next strike flies by without a flinch from the batter. Taking the first strike may be an acceptable strategy, but we will see how a probability tree confirms the fevered outrage of fans when our batter took that second strike. We will see that for anyone understanding the probabilities there is no joy in the batter having two strikes!

To dig into the probabilities, let's look at an average batter's plate appearance after letting two strikes pass. Each plate appearance in a baseball game ends on one of twelve possible pitch counts: $(0 - 0)$, $(1 - 0)$, ..., $(3 - 2)$, where the first number in the ordered pair represents the number of balls the pitcher has thrown to that batter, and the second represents the number of strikes thrown. Let's assume that nothing out-of-the-ordinary happens during the at bat to cause it to terminate, such as a base-runner getting thrown out trying to steal, the pitcher balks, the batter is hit by the pitch, etc. In other words, the plate appearance will end in either a hit, out, or walk. Of at bats that terminate with the ball being put into play, there is a chance that this event will result in a hit. Table 14.2 gives the percentage of balls put into play during each of the twelve possible terminal counts that resulted in a hit accumulated from the 2009–2011 Major League Baseball seasons.

This data was obtained by downloading pitch-by-pitch game data available at www.retrosheet.org. At bats were then categorized based on the result and the final pitch count. It is clear from the chart that when a count reaches two strikes, the advantage swings heavily toward the pitcher. This could be due to the batter having to swing at bad pitches to protect the strike zone or the pitcher fooling the batter with an unexpected pitch. Regardless of the reason, once the count goes to two strikes, the batter clearly, and rightfully, feels the pressure.

Let's take a look at an at bat that currently has a (0–2) count, and use the data in Table 14.2 to see what percentage of the time the at bat ends in the batter's favor. For simplicity, let's assume that with each pitch the batter will swing 75% of the time (to protect the plate) and that the pitcher will throw a strike 60% of the time regardless of the pitch count. We will also assume that there will be no foul balls on any swing. In this situation, we can easily draw a probability tree, as in Figure 14.5, outlining the decisions

Table 14.2: Batting averages per terminal pitch count.

Count	Hits	Total	Average
(0-0)	19544	57870	0.337722
(1-0)	13308	38868	0.34239
(2-0)	4994	13867	0.360136
(3-0)	292	751	0.388815
(0-1)	15567	48463	0.321214
(1-1)	15655	47623	0.328728
(2-1)	10117	29700	0.34064
(3-1)	4408	12462	0.353715
(0-2)	6928	44547	0.155521
(1-2)	13093	77842	0.1682
(2-2)	13945	74095	0.188204
(3-2)	11151	48929	0.227902

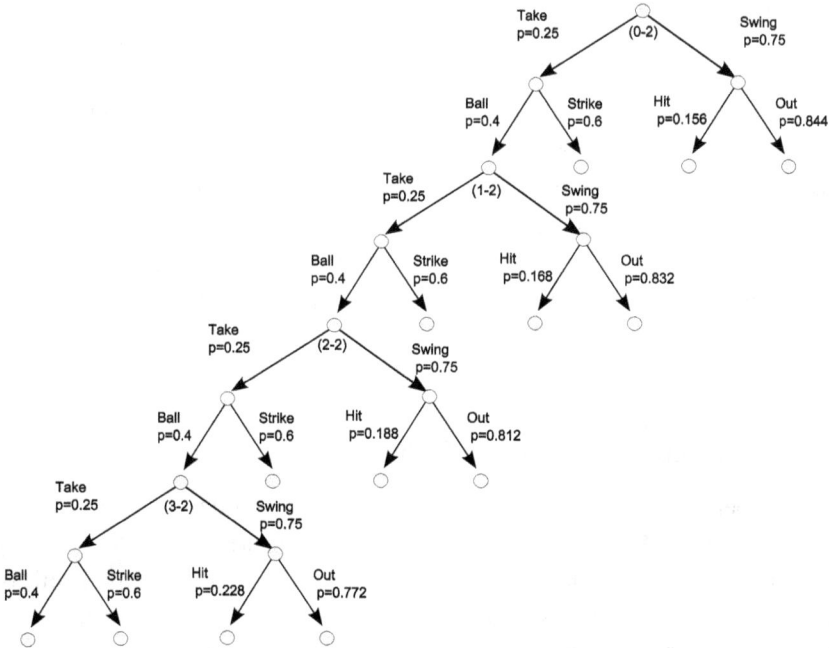

Figure 14.5: A probability tree starting from an (0–2) count.

in the problem, the results of those decisions, and the probability of each decision.

It is then easy to calculate the probability of each of our targeted outcomes by identifying the paths in the tree that terminate with each of three results. There are four paths in the tree leading to a hit. Then,

$$\begin{aligned} \Pr(\text{Hit}) = {}& \Pr(\text{Hit on (0-2)}) + \Pr(\text{Hit on (1-2)}) + \Pr(\text{Hit on (2-2)}) \\ & + \Pr(\text{Hit on (3-2)}), \\ = {}& (0.75)(0.1555) + (0.25)(0.4)(0.75)(0.1682) \\ & + (0.25)^2(0.4)^2(0.75)(0.1882) + (0.25)^3(0.4)^3(0.75)(0.2279), \\ = {}& 0.1312. \end{aligned}$$

There is only one path to a walk in the tree, namely the batter must take four balls in a row. Hence, $\Pr(\text{Walk}) = (0.25)^4(0.4)^4 = 0.0001$. All other paths in the tree lead to an out. Thus $\Pr(\text{Out}) = 1 - \Pr(\text{Hit}) - \Pr(\text{Walk}) = 0.8687$. When a batter reaches a (0–2) count, the probability of a successful plate appearance reduces to a woeful 13%.

If you find your team down to one more out in the ninth inning with a batter facing a (0–2) count, you can rightfully feel a sense of dread. Yet, like the fans in Mudville, you may just sense that your batter just won't "let that ball go by again". We did, after all, find a probability which gives the tendency over many such at bats. As infielder Toby Harrah stated on statistics and bikinis, "They show a lot, but not everything." So, feel free to jump out of your seat and root for the home team as if they don't win, it's always a shame.

14.4. Probable Super Bowl Score

The Super Bowl is about as close to a sporting holiday as it gets in the United States. Millions of dollars are spent to ensure that those watching the Super Bowl keep their eyes on the screen. There is the game itself, of course. Then, here are also the Super Bowl commercials. From Matthew Broderick's 2012 remake of Ferris Bueller's Day Off to the outlandish game of H-O-R-S-E played by Michael Jordan and Larry Bird in 1993, it is clear that Super Bowl ads are memorable and attention grabbing.

In spite of these efforts, there are still some things that will compete with the on-screen action on a Super Bowl Sunday. Large spreads of snacks and interesting conversations at parties have the potential to distract viewers from the game. What happens if an interruption causes you to miss a series of plays that led to scoring? Fortunately, you can use a little math to help you get caught up on the action.

To see how, suppose you missed the entire first half of the 2013 AFC Championship Game between the New England Patriots and the Baltimore Ravens. When you joined the game at the start of the third quarter, you see that the score is 13-7 with New England on top. You say to yourself, "I wonder how New England scored those points." There are a couple of natural first guesses:

1. $7 + 3 + 3$: a touchdown, an extra point, and two field goals.
2. $7 + 6$: two touchdowns, one extra point, and one missed extra point.

There are some other possibilities as well. Among them are

3. $8+3+2$: a touchdown, a two-point conversion, a field goal, and a safety.
4. $7 + 2 + 2 + 2$: a touchdown, an extra point, and three safeties.
5. $3 + 2 + 2 + 2 + 2 + 2$: a field goal and five safeties.

Football intuition would tell you that some of these scoring scenarios seem much more probable than others. Let's quantify this intuition by considering the average number of times each of the following events occurs in an NFL game (where the average is computed using data gathered from the 2012–2013 season and its previous two seasons).

Scoring event	Average number per game
Touchdown	5.10
Extra point (EP)	4.84
2 point conversion (2PC)	0.12
Failed EP or 2PC	0.15
Field goal	3.31
Safety	0.05

Now, each team in an NFL game has, on average, 12 possessions per game — a total of 24 possessions in all. The chart above indicates that there are approximately five touchdowns and three field goals in a game (on average). This means that roughly five out of 24 possessions (about one fifth) result in touchdowns and three out of 24 possessions (one eighth) result in field goals. We also note that since a safety occurs in roughly five out of every 100 games (or 2400 possessions), the probability of a safety occurring in a given game is 5/2400 (or 1/480). A similar reasoning shows that the probability of a two-point conversion occurring is about 12/2400 (or 1/200).

How can we use these probabilities to discern the likelihood of some of the previously listed scoring scenarios? As is often helpful in mathematics, we consider a simpler problem first in order to get a better sense of the underlying situation.

Suppose we are playing a game in which we select gumballs, one at a time, from a bucket that contains five white, 10 red, and 15 blue gumballs. Once we choose a gumball, we note the color and then put it back into the bucket. If we were to do this five times, what is the probability that our five choices would have included two reds, two whites, and one blue?

Since 5 of the 30 gumballs in the bucket are white, each selection has a 5/30 (or 1/6) chance of being white. Similarly, each selection has a 1/3 chance of being red and a 1/2 chance of being blue. Since the selections are independent of one another, the probability that our five choices were, in order, red-red-white-white-blue is $(1/3)(1/3)(1/6)(1/6)(1/2) = 1/648$. There are other orderings (30 in all) that would produce the same color totals, and each such ordering yields the same probability (1/648). This means that the probability of choosing two reds, two whites, and one blue in our five choices is 30(1/648), or about 4.6%.

Using a similar reasoning, we can conclude that selecting three blue, two red, and no white gumballs has a probability of 13.9% — roughly three times more likely than our earlier situation. This makes sense, since we see that the bucket contains more blue gumballs than any other color.

What does this mean for our football scores? What are the probabilities of the different scoring combinations that add up to 13 points? Just as we were 3 times more likely to select a blue gumball than a white one, a field goal is 60 times more likely than a safety and a touchdown is 96 times more likely than a safety. Using these probabilities, we find that the probability of getting 13 points with a touchdown, extra point, and two field goals is around 6.2%. The probability of getting 13 points with two touchdowns, one extra point (and one missed extra point) is 0.66%. This means that scenario (1) above is almost 10 times more likely than scenario (2). The striking difference here is due to the fact that missed extra points are rare. Safeties are rare as well, and we can calculate scenario (1) to be almost 2.5 million times more likely than scenario (4) with a touchdown, an extra point, and three safeties. So, if you see a score of 13 at halftime and don't think about the possibility of three safeties, that's certainly understandable!

What about other possible score totals? As you would imagine, as the totals increase, so do the numbers of possible ways to achieve the scores. For instance, there are 31 different ways to get to 27 points and there are 41 ways to get to 30 points.

How about the 1916 game between Georgia Tech and Cumberland? Georgia Tech scored 222 points to Cumberland's 0. There are 8,162 ways that a score of 222 can be obtained. Many of these scenarios are about as likely as the far-fetched shots made by Jordan and Bird in the 1993 commercial (for instance, 111 safeties) — but they are possibilities nonetheless.

We can simplify the problem, however. From the scoreboard in Figure 14.6, we see Georgia Tech scored 63 points in the first and second quarters and 54 and 42 points in the third and fourth quarters. There are 247 ways to score 63 points. Similarly, there are 168 and 90 ways to score 54 points and 42 points, respectively. So, if we consider how many points were scored in each quarter of the game, we find that there are $247 + 247 + 168 + 90 = 752$ different ways for Georgia Tech to reach their final score in the historic game. If we assume no safeties, then there are only 80!

In the immortal words of Ferris Bueller, "Life moves pretty fast. If you don't stop and look around once in a while, you could miss it." Well, if

Figure 14.6: In 1916, Georgia Tech and Cumberland played a historic lopsided game. So much so, you might even have a double take on the scoreboard.

you miss any part of a Super Bowl, don't just hit the rewind button. Use the opportunity to explore the possibilities with math! Even though your couch-sitting body isn't getting any exercise, you can give your brain a good mental workout!

Spoiler Alert

Here is our answer to our earlier basketball question. Again, Player A shoots 25% from the free throw line while Player B shoots 75%. Both players get two shots after being fouled. For either player to score 1 point, then either the first shot is made and the second missed or the first shot is missed and the second made. The probability that Player A makes the first shot and misses the second is $(0.25)(0.75) = 0.1875$. Note, however, that this is the same computation that we'd make for Player B missing the first shot and making the second. Similarly, the probability of Player A missing the first shot and making the second is $(0.75)(0.25) = 0.1875$, which is the same computation for Player B making the first shot and missing the second. So, Player A will score exactly one point 36.5% of the time, which is precisely the same probability of Player B scoring exactly one point. Now, this could seem like a bit of a trick question since the answer is both Players A and B. This contains a good lesson about mathematical language. The question asked "Player A or Player B?". In this case, the "or" is really intended to be an *inclusive or*, meaning A or B or both. If we meant A or B but not both, then mathematically, we'd write "Player A or Player B", which is called an *exclusive or*.

Chapter 15

Robot Reporters

In times past, sports reporters sat in the press boxes or sidelines taking notes for the stories they'd write about the game. Today, the reporters may still be in the stands but, in some cases, the "reporters" are the fans sharing their stories on social media or someone at the press table recording play by play. In this chapter we'll see how computers can compose text. As such, we may some day see a Pulitzer Prize for reporting go to a computer. A key is how the Internet and social media give up-to-date information on our world. News outlets are using social media to help write stories on games in which no reporter attended. In fact, some such stories are automatically written by a computer.

15.1. Automated Storytelling

A host of sports information exists on the web. Modern news outlets are harnessing that information to cover events automatically via computers. Computational reporting has already been writing sports reports and business news. For example, computers are supplying daily market reports for *Forbes* as well as sports reports for the Big Ten sports network. In [1], the following automated example is given from online baseball stats:

> "Tuesday was a great day for W Roberts, as the junior pitcher threw a perfect game to carry Virginia to a 2–0 victory over George Washington at Davenport Field. Twenty-seven Colonials came to the plate and the Virginia pitcher vanquished them all, pitching a perfect game. He struck out 10 batters while recording his momentous feat. Tom Gately came up short on the rubber for the Colonials, recording a loss. He went three innings, walked two, struck out one and allowed

two runs. The Cavaliers went up for good in the fourth, scoring two runs on a fielder's choice and a balk."

This work has become so developed that automated storytelling is a field. In 2017, *The Washington Post* used its in-house automated storytelling technology to cover each week's high school football games in the Washington, D.C. area [62]. How well can these be written? In 2014, a Swedish media professor, Christer Clerwall, compared reports written by human and computers via a blind study. Readers tended to find human sports journalists' stories more accessible and enjoyable. Computers, however, wrote stories that tended to be seen as slightly more informative and trustworthy.

With advances in machine learning, we are likely to see robots doing a variety of our news reporting! So, let's step into computational text generation.

15.2. Going Ape on a Typewriter

To explore how to write automated stories, let's be really random and go ape on a typewriter (Figure 15.1). Suppose a chimpanzee begins typing on a computer's keyboard or typewriter. What text will result? Hard to say, really.

Figure 15.1: Even one ape typing along at random is guaranteed to eventually type out the entire contents of *War and Peace*, although we may have to replace our apes as the timeline for such an event may be longer than the life of quite a few apes.

Let's assume that the ape is equally likely to hit any lowercase letter or the space bar. So, the chimp isn't fascinated with the letter "y", for instance, any more than the letter "u". If we put an army of chimpanzees at word processors, eventually one will bash out the exact contents of your favorite song.

Let's think about that! We are *guaranteed* that an army of chimpanzees can eventually produce the exact lyrics of your favorite song. Granted, the army of apes might be larger than the current population of apes on the Earth and might take longer than the age of the universe to produce that lettering. Even so, at any moment, you could look down and see that any given ape had typed something like:

oognqskj   bxyvchjsdtcrnvtyyjsfaunmyqqxvpegxane   tnmbsbbocwwtero rvmlwcdekwvbkolrqhla eckfnjzybthlozl isrosred eapxsfjm exrkflndpgkpghqhw tjpcyxwhqalieelcqmssrabiorlwt oicqvlchehlomxn

So, our army of chimps could pound away. Periodically, we could check if one had composed the lyrics of interest. While such lyrics are guaranteed to appear, it is a most inefficient and impractical method to write them. Let's try another approach that will generate random text but not necessarily as random as that of our chimp friends.

First, remember that most events occur in context. Suppose that we wanted to randomly generate a year's worth of Fahrenheit temperature data. A series of 365 random integers chosen uniformly between 0 and 100 wouldn't fool the average observer. If it is 85 degrees today, it probably won't be 5 degrees tomorrow.

The same is true of English words: if the current letter is a "q", then the next letter is quite likely to be a "u". A random text generator can make more interesting text by selecting each letter using the probability it follows the preceding letter. We could, therefore, read a sample text and count how many times every letter follows an "a", how many times they follow a "b", and so on for each letter of the alphabet. When we write the random text, we produce the next letter as a random function of the current letter.

Let's look at a small example. Let our sample text be the title of an article written by Robert R. Coveyou,

"The generation of random numbers is too important to be left to chance."

We'll consider this as being written in lowercase and remove punctuation. Suppose we start our random text with an "a", which appears four times in the sample text. Looking at the quote by Coveyou, we see that an "a" is followed three times by the letter "n" and once by the letter "t". Generating the random text, we will assume there is a 75% chance that an "n" will follow an "a" and a 25% chance that a "t" will follow an "a". If we produce an entire text this way, also choosing the first letter randomly, the random text

```
hers ne is bef is chen be le
```

could be generated.

Let's try a larger sample text by inputting the first chapter of the *Adventures of Huckleberry Finn* by Mark Twain. The following text resulted:

> I me bawithip bup by g akealdaitr. memugoin wheeday t cofome ban soromeal basin ssath, angher co I g shetidenousomenoishet gen an mixes st t an't. whed, ry, do therst bul s: Shatrois ad imarered we?"men mes, YOU whe r boon Do; tr, in've. hinge alan thoutr bacong ossmy t judn h ctout he st warerain, kis or d odndl t g, bo sherithe sabbomuthit p ch misceede ne- thes s utapit gheth wndnd w idet f tors Sheskere n l, thing cheaboouthe be; ithay wica cheand g un't bun tanomme ny aido I che me, ais arshin, she ad f ce me sedoderuld thed Mar whancemakeven'thet wan Hus Mang tusttome theancr abo. ithenthe; me.

CHALLENGE: READ THE FOLLOWING

```
The lertets in tihs txet are pclead at roandm
eexcpt for the frist and lsat ltetres of ecah
wrod wcihh hvae the smae pnelemact. Can you
raed it? The biran is pruwloef!
```

Can you tell how the text above is created? Note how simply having the letters that appear in a word, even if they are scrambled, but having the first and last letters in place, allows us to recognize the word. No wonder we can overlook spelling mistakes in our writing!

In such a way, we can set up a process to emulate language. If one letter is printed, we can decide at random what letter will follow if we know the probability of every other letter following. If we consider this one letter at a time, then we are creating what is known as order-1 text. This often does not produce very recognizable text as the earlier example displays.

Let's extend this idea to longer sequences of letters. Let's construct order-2 text by taking 2 letters at a time. Now, we write a letter by noting the two letters preceding it (a letter pair is often called a digram). The digram "th", for instance, is often followed in English by the vowels "a", "e", "i", "o", "u", and "y", less frequently by "r" and "w", and rarely by other letters. Such an idea can also be extended to create order-*n* text. We'll create an order-2 text by again using the Mark Twain text.

> Nown maidn't a shen haver he sting cough, ith liked bace. But I wit out mong and aidow and, able liped all off of abled ust ned to in thaver. Now Dousell swer. I nes whime whe wou hown maidong a he I wout mad smoneverying tice he was eaticul over able told Mis — so luck. a cry. Then and saight;" and untell saidn't grouread toger, the up to wheatterybougladerried then toll a liver to it ing ithat she vilipery;" dowen a coulas — she hathe thad, a dong ho-whe thee jushey stich so uppers, al all ou hat, a sed re gerryind sup and a sigh all a thound "Dout I hen that sheang widn't no but gaidown in all and sair the go ing, yould. Aund glas

15.3. Computational Shakespeare

We can also generate random text at the word level. Like the text of the apes, we could randomly generate words from a dictionary. This would result again in largely meaningless results. A slightly better approach reads a document, counts the number of times each word appears, and then produces text by printing words according to their probabilities in the source text.

Let's take the approach we took with letters and take into account the preceding word group in order to generate the next word group. This is the same process as with the letters we previously viewed. Order-1 and order-2 text produce the current word as a random function of the previous word and two words, respectively. Order-1 text is generally almost readable. Order-3 text tends to contain long phrases from the original text. Order-2 is, indeed, readable but does not contain those long passages from the original text. It sounds original!

As an example, suppose we input the first scene of the first act of *Romeo and Juliet* by William Shakespeare. The following order-2 text output is generated from one run of this algorithm:

ACT I

PROLOGUE

Two households, both alike in dignity,
In fair Verona, where we lay our scene,
From ancient grudge break to new mutiny,
Where civil blood makes civil hands unclean.
From forth the golden window of the east,
A troubled mind drave me to forget.

BENVOLIO: I'll pay that doctrine, or else die in debt. (Exeunt)

MONTAGUE and LADY MONTAGUE: O, where is Romeo? saw you him to-day? Right glad I am for you: I serve as good a man as you.

ABRAHAM: No better.

SAMPSON: Well, sir.

GREGORY: Do you quarrel, sir?

ABRAHAM: Quarrel sir! no, sir.

SAMPSON: If you do, sir, I am a pretty piece of flesh.

GREGORY: 'Tis well thou art not quickly moved to strike.

SAMPSON: A dog of that house shall move me to forget.

BENVOLIO: I'll pay that doctrine, or else die in debt.

Exeunt all but MONTAGUE, LADY MONTAGUE, and BENVOLIO

Sound dramatic and Shakespearean? In some cases, yes and other sentences no. Let's move from a computational playwright who will pen comedies for the ages to creating a computational poet.

15.4. Mathematical Love Poem

How many times have we seen someone in a movie or on television ruining a lovely moment due to a loss of romantic words? If only such instances could mirror the triumphant story of Cyrano de Bergerac who whispered heart-melting sentiments to a suitor who otherwise would have been at a loss for words.

Can we harness the power of math and randomness to create poetry that might pierce the heart like one of Cupid's arrows? Can a computer realize the following quotation of Plato: "At the touch of love everyone becomes a poet."

We'll use the ideas of the last section. So, we need to input text. It should be clear that the type of text we input largely influences the style of writing that will be randomly produced. Many options exist – love poems, sentiments on greeting cards, and love songs.

Let's move in the direction of love songs and see what emerges. We'll input lyrics from the Elton John songs "Love Letters", "Love is Worth Waiting for", "Love is Here to Stay", "Love is a Cannibal", "Love Her Like Me", and "Love Adventure".

Now, the ever-important question emerges — *How do we do?* Can our computational poet create something useful? You judge. On an initial run of the program, the following emerges:

Love is painted grey
They did it for a year
but ever and your heart
But, oh my eyes
and the dog who follows
the Rockies may tumble
They're only made of love

Not going to be effective for the love of your life? Run the program again! A second run produces another romantic sentiment:

Love is here to stay
Together we're going a long, long way
In time the baby and the sky
And love her conscience clean
You can charm her
Keep her conscience clean
You can have her when she's wild
Show a flag on the week
You can never, never
love eats love
Love letters straight from your old money
Dance around the night

Still not your desired words? You don't imagine your love's knees quivering? If not, a few options remain. First, you could add more input text as this creates more options for the output. Alternatively, you could simply run the program many more times until you find something suitable. Also, note that the outputted text was much longer and the listed outputs are simply the first few lines. Another option would be to pick and choose from the outputted text. For the purposes of this book, only the first few lines of text were chosen to demonstrate what can emerge from the most straightforward application of our computational love poet.

You may also find that you're not a fan of lyrics to Elton John songs. In fact, it may be that the songs even in their original version don't speak to you. If so, you could try inputting text more dear to your heart and hopefully more dear to your dearest heart! In the end, the computational bits and bytes may produce random text that captures exactly what you were wanting to say!

RANDOM RESEARCH RESULTS

Academics often receive spam emails requesting submission of their scientific papers to conferences. In 2005, MIT students Jeremy Stribling, Daniel Aguayo, and Maxwell Krohn wrote a program to generate a nonsense scientific paper. The program produced the paper "Rooter: a methodology for the typical unification of access points and redundancy". The group submitted the paper to the 2005 World Multi-Conference on Systemics, Cybernetics and Informatics. The conference organizers claim that the paper was sent to human reviewers. While no comment was made on the paper, it was accepted as a scientific paper. They were asked to attend the conference (with its $390 registration fee) and give a talk. The students shared the news of the paper's acceptance with the larger academic community and in time the conference banned the paper [46]. Interested in making your own random scientific paper and reviewing its content? Search the internet and you'll find programs that will create "research papers", for instance in computer science. (Note that generating the paper and reading it is recommended, submitting the paper to a conference is not!)

Chapter 16

It's Only a Game

Let's get into the game by playing games. We'll begin with a game that overlaps our topic of Monte Carlo simulation for Chapter 9. Then, we'll see how math can create unbeatable games and also teach us mathematical ideas.

16.1. Die-ing to Play Baseball

Let's play baseball but with dice. We'll use the 2016 data for Major League Baseball as given at baseball-reference.com. There were 165,561 at bats resulting in 27539 singles, 8254 doubles, 873 triples and 5610 home runs. Converting these to probabilities, in 2016 for each at bat, there was a 16.6% chance of having a single, a 5% chance of having a single, a 0.5% chance of a triple, and a 3.4% chance of a home run.

We use these probabilities to create our dice combinations that relatively approximate the percentages. You roll one die. If you roll a 1, you have a single. Note, that this occurs $1/6 = 0.166$ of the time. A double will occur if you roll a 2 and then roll again and get either a 1 or a 2. The probability of this is $2/36 = 0.055$, since rolling a 2 and a 1 or a 2 and a 2 are two combinations among the 36 ways you can roll a die twice. If your first roll is a 3, then you have a chance of a triple. But you must roll three straight threes in order to get to third base, which corresponds to a probability of $1/216 \approx 0.46\%$ since rolling three straight threes equates to one option among the $6^3 = 216$ ways to roll a die three times. To hit a homer, you can roll a 4 or a 5 on your first roll. If you roll a 4, then you need another 4 to hit a home run. If you rolled a 5, you need to roll two more fives to hit a homer. This probability equals $7/216 = 0.034$. We'll think of it as rolling a die three times. One of the options is to roll three fives, which is one of the 216 options of rolling a die three times. You can also roll a 4 and a 4 and anything on the third roll; there are six ways to do this, giving us seven total ways to hit a home run. So, pick up a die and play ball!

Let's do the same thing but for specific teams. Take the Chicago Cubs who won the 2016 World Series. The Cubs, as a ball team, had 5503 at bats. They had 887 singles, 293 doubles, 30 triples, and 199 home runs. This corresponds to a probability of 0.161 for a single, 0.053 for a double, 0.0054 for a triple, and 0.0362 for a home run. So, our rule for a single can stay the same. The Cubs had a lower than average number of doubles. Our current rule equates to rolling a 2 and then a 1 or a 2 and anything on the last roll. There are 12 ways to do this with rolling a die three times. We only want 11 ways for the Cubs. This can be fixed by saying that you get a double if you roll a 2, get a 1 or a 2 on the second roll and get anything except 6 on the third roll. The Cubs had a higher home run percentage than MLB. Our current rule has a $7/216$ probability of occurring. Rather than 7 outcomes yielding a home run, we now want 8 to better approximate the Cubs' performance in 2016. So, our rule can become rolling a 4-4, 5-5-5, or 6-6-6.

You can also look at individual players and create combinations of rolls that approximate their batting average or the average number of singles, double, triples and home runs they hit. In this way, you can simulate an entire game. The dice approximate the percentages. You could also use a random number generator to better approximate the actual averages accumulated in 2016. You simply need the data and a die and you're ready to get in the game of baseball.

16.2.   A Winning Fib

Let's try another game. You could play with pebbles, candies, or sticks. We'll play with clothespins hanging on a clothesline as seen in Figure 16.1. We'll take turns taking between 1 and 3 clothespins off the line. Whoever takes the last clothespin loses. You can go first. You think big and take 3 of the 17 clothespins. In an attempt to have the game last longer, I'll only take 1. You decide to go big or go home and take 3 again. I'll stick with our

Figure 16.1: Learn the rules of the game and play. If you take the last clothespin, you lose!

pattern and take 1. You decide to slow things down and take 1. I liked your pattern so I'll take 3. So, there are five remaining. Can you win? If you take 1, I'll take 3, and there is 1 remaining that you must take. If you take 2, I'll take 2 and again, you'll be left taking the remaining clothespin. And yes, taking just 1 leaves you with the same fate.

This can easily seem rather contrived as I'm determining your choices, which forced us down to 5 clothespins guaranteeing my win. It turns out what guaranteed my win was starting the game with 17 clothespins. You go first and if you take 3, I take 1. If you take 2, I take 2. If you take 1, I'll take 3. Said another way, I make sure that 4 clothespins are taken after you and then I have gone. We started with 17 and $17 = 4(4) + 1$. Dividing 17 by 4 and having a remainder of 1 guarantees my win!

In our baseball game with dice, the randomness of rolling dice plays into the game so you don't know what will happen. This game is deterministic. If you play the way just outlined, you know who is going to win. The issue with the version we just played is that someone can notice what you're doing, count the original number of clothespins and figure out what's happening.

Let's obscure the winning strategy a bit but still play this style of game, which is often called nim. The rules differ this time.

On the first move, a player is not allowed to take all of the clothespins, and on each subsequent move, the number of clothespins removed can be any number that is at most twice the previous move. The player who takes the last clothespin wins. Again, go ahead and go first. Suppose you take 3. Now, I can take up to double what you took, which would be 6 clothespins. Like the last game, I'll simply take 1. You can only take 1 or 2, so you take 2. This time, I get bigger and take 3. This leaves us 8. If you take 3, I can win by taking the last 5. So, you want to only take 1 or 2. You take 1. There are 7. I take 2, leaving 5. You clearly don't want to take 2, as I'll take the remaining 3. So, you take 1, leaving 4. I'll take 1. There are 3 and you can take 1 or 2 clothespins. If you take 1, I can take the remaining 2. If you take 2, I can take the remaining 1.

Maybe we should go back to the point we had 8 and you could have taken 1 or 2. I let you take 1. Instead, let's have you take 2. I will take 1, leaving 5. You can't take 2 or I take the remaining 3, and we already saw that taking 1 leads to a loss. Maybe we should go back more and choose other numbers. This wouldn't help, as we again have a version of nim in which the second player can always win.

The strategy leans on the Fibonacci sequence, which as we saw in Section 1.6 are numbers in the sequence 0, 1, 1, 2, 3, 5, 8, 13, ... Any following term in the sequence equals the sum of the previous two. The winning strategy relies on Zeckendorf's theorem, named after Belgian mathematician Edouard Zeckendorf. In particular, every positive integer can be represented uniquely as the sum of one or more distinct non-consecutive Fibonacci numbers. The strategy is for me to find the non-consecutive Fibonacci numbers (also known as the Zeckendorf representation) that sum to the number of current clothespins. The number of clothespins I take equals the smallest number in that sum.

Again, on my first turn since you took 3 of the 17, there are $14 = 13 + 1$ clothespins. So, I took 1. You then took 2, leaving $11 = 8 + 3$. So, I took 3, leaving 5. You took 1, leaving 4 and $4 = 3 + 1$, indicating I should take 1 and, as we saw, this leads to a winning game for me.

There's only one catch. Pick the number of clothespins such that it isn't a Fibonacci number. If it is, the first player can win!

16.3. Dodgeball on Paper

Let's play dodgeball with pen and pencil as described in *Coincidences, Chaos, and All That Math Jazz: Making Light of Weighty Ideas* by Burger and Starbird [10]. There are two game boards as seen in Table 16.1 at the end of the chapter. The game begins with Player 1 filling in the first row of empty boxes of Player 1's game board with X's and O's. Then, Player 2, known as the Dodger, places either an X or an O in the first box in the Dodger's game board. Player 1 then writes a sequence of X's and O's in the second row of empty boxes in Player 1's game board. Then, the Dodger places an X or an O in the second box of the Dodger's game board. The game continues in this fashion until all six rows of Player 1's game board are filled with X's and O's and the Dodger's row is also full. Player 1 wins if any of the rows in Player 1's game board exactly matches the Dodger's row of X's and O's. If they don't match, the Dodger wins. Can you devise a strategy that will guarantee a win for either player? Hint: there is one! Play the game a few times and see if you can figure it out. Grab a friend and work together.

> **Don't continue to read until you are ready to know a winning strategy. Remember, once you read it, you won't have the chance to discover it!**

To win, you want to be the Dodger. Here's the strategy. When Player 1 fills the first row, look at the first letter. If it's an O, write an X. If it's an X, write an O. Now, when Player 1 fills the second row, look at the second box. If it's an O, write an X. If it's an X, write an O. See the pattern? Said mathematically, when Player 1 fills the ith row with X's and O's, look at the ith box in that row. The Dodger will write an X if that box contains an O and an O if it's an X. This guarantees that the Dodger will win. When the game ends, the Dodger's row cannot exactly match any row of Player 1. Why? Take a random row on Player 1's game board. It must differ by at least 1 box. When Player 1 wrote down that row of X's and O's, the Dodger ensured this to be the case. We'll soon see that this is infinite ramifications, literally!

Try playing and when you've mastered the strategy, read the next section as we'll play on an infinite board.

16.4. Dodging Infinity

In *Auguries of Innocence*, William Blake beckons the reader to "...hold infinity in the palm of your hand." While we can't see the infinite, our game of Dodgeball for consistency can help us explore it.

In Disney's movie *Toy Story*, Buzz Lightyear exclaims, "To infinity and beyond." What could possibly be beyond infinity? If you combine the positive numbers with the negative numbers, and toss in 0, you have the set of all real numbers. In terms of infinity, we've essentially taken infinity + infinity = infinity.

Infinity measures the size of a set like the set of natural numbers, which are the numbers 1, 2, 3, 4, and so on. Two sets have the same size if you can

Table 16.1: Game boards for Dodgeball for consistency that can teach us concepts of infinity.

Player 1's game board

1					
2					
3					
4					
5					
6					

The Dodger's (Player 2's) game board

1	2	3	4	5	6

describe an exact pairing of their elements. In the picture below, we have three cents since we can line up our pennies, as shown, with the numbers 1, 2, and 3. Each of these numbers pairs with exactly one penny.

1   2   3

To think about the infinite, we'll visit, as we did in Chapter 9, the Math Pub, where all things mathematical but not necessarily physical can occur. The Pub services the Mega-Motel, which offers infinitely many rooms and is booked solid. You walk up and ask for a room. The attendant at the front desk states, "Welcome, here is your key for room 1." Then, over the intercom, everyone is asked to move down one room. Does everyone fit? If not, who doesn't? Think about it. Everyone knows where to go. This helps us see that the sets $\{0, 1, 2, 3, \ldots\}$ and $\{1, 2, 3, \ldots\}$ have the same size. So, even if you had arrived in a bus of 100 people, the motel could accommodate everyone. Loosely, this isn't all that surprising since putting you in room 1 can be thought of as infinity $+1 =$ infinity.

Is the only thing beyond infinity the infinite itself? In a sense, yes, but possibly not in the way one might expect. Consider what it means for two sets not to have the same size. We have two rows of pennies below. If we think of them as sets, we can clearly see that the set of pennies in the first row is not the same size as the set of pennies in the second row. In particular, these two sets are not the same size since any pairing of the sets will always leave a penny in the second row unpaired with a penny in the first row.

Let's return to our Dodgeball for consistency winning strategy. Our rows had six boxes. If we think of the game as having infinitely many boxes in each row, we can see something about infinity that can be astounding when first learned. In the late 19th century, Georg Cantor analyzed the infinite with a strategy much like our Dodgeball winning strategy. In particular, he looked at the set of real numbers between 0 and 1 and the set of natural numbers. Let's assume the sets are the same size, which makes sense as infinity is infinity. This means we can define an exact pairing between all the elements of the sets. Assume this exists. Then, we can't produce a real number between 0 and 1 that isn't paired with one of our natural numbers. However, the Dodger knows how to produce such a number!

Suppose our pairing is

$$
\begin{aligned}
1 &\longrightarrow 0.\underline{7}65242\ldots \\
2 &\longrightarrow 0.6\underline{3}2314\ldots \\
3 &\longrightarrow 0.13\underline{4}210\ldots \\
4 &\longrightarrow 0.024\underline{5}25\ldots \\
5 &\longrightarrow 0.3029\underline{3}1\ldots \\
&\qquad\qquad \vdots
\end{aligned}
$$

The Dodger will choose the number 0.30330.... Why? To choose the first decimal digit, the Dodger looks at the first decimal digit of 0.765242... since it is paired with the number 1. Since the first decimal digit is not equal to 3, the Dodger's number will have 3 as its first decimal digit. To choose the second decimal digit, the Dodger looks at the second decimal digit of 0.632314... since it is paired with the number 2. Since the second decimal digit is a 3, the Dodger's number will have 0 as its second decimal digit. To choose the third decimal digit, the Dodger looks at the third decimal digit of 0.134210... since it is paired with the number 3. Since the third decimal digit is not a 3, the Dodger's number will have 3 as its third decimal digit. This pattern would continue for all the digits in the list producing a number that cannot be contained in the pairing. Note that this is Dodgeball for consistency on an infinite board. Regardless of a pairing between the set of natural numbers and the set of real numbers between 0 and 1, the Dodger can create a decimal number not in the pairing. This shows that while both sets are infinite in size, one is bigger than the other! Indeed, there is more than one size of infinity and Dodgeball enabled us to prove it.

So, maybe Buzz Lightyear was right. We can go to infinity and beyond — to another sized infinity. What lies beyond that size? There is, indeed, another sized infinity. How? Look it up in a book or on the Internet. You'll be reading another example of how mathematics allows us to study abstract ideas in ways that might initially defy intuition. Does more than one size of infinity do that for you? It might. It did for colleagues of Cantor. In fact, Henri Poincaré, a leading mathematician of the day, called Cantor's ideas a "disease". In time, Cantor's work was embraced. Accomplished mathematician David Hilbert, who originated the story of an infinite hotel, stated, "No one will drive us from the paradise which Cantor created for us."

Mathematics continually pushes the boundaries of its knowledge ... in a way, to infinity and beyond.

Chapter 17

Playing in a Higher Dimension

Viewing ideas through the lens of mathematics enables us to examine the unseen. Having just viewed the infinite via mathematics, let's imagine another seemingly unimaginable idea, the 4th dimension, by examining the relationship between the 2nd and 3rd dimensions. This will help us to gain perspective on moving from 3D to 4D.

17.1. Managing the MLB in 13 Dimensions

First, how can we even see in the fourth dimension? Let's think bigger and graph in 13D! We'll use Chernoff faces, an idea created by Herman Chernoff, where data is displayed in the shape of a human face. It can help us to think of it as a Mr. Potato Head for data analysts. Let's see how it works by visualizing 3D data in 2D. Our data varies between 0 and 10. We'll have three data points, and they will adjust the shape of the head, the size of the eyes, and the size of the nose. To help us to get oriented, the point $(0, 0, 0)$ corresponds to the image in Figure 17.1(a). In contrast, the point $(10, 10, 10)$ corresponds to the image in Figure 17.1(b). The points $(5, 5, 5)$ and $(2, 7, 10)$ give us the Chernoff faces in Figures 17.1(c) and 17.1(d), respectively.

Plotting 3D data in 2D isn't all that new. You've essentially done this when you have drawn a cube on a piece of paper. But we can add more features and move beyond the 3D. For example, let's now have our data vary the shape of the head, space between the eyes, size of the eyes, shape of the ellipse, pupil size, slant of the eyebrows, size of the nose, size of the mouth,

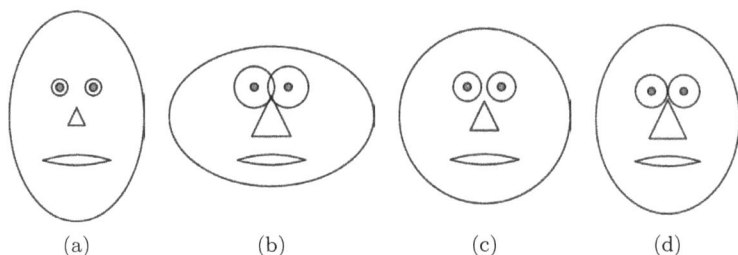

(a)   (b)   (c)   (d)

Figure 17.1: Chernoff faces enable us to visualize higher dimensional data in two dimensions. Adapted from [64].

Figure 17.2: Chernoff faces visualizing data in the 10th dimension.

shape of the mouth, and opening of the mouth. If you were counting, that puts us in the 10th dimension! Let's create random points in the tenth dimension with each coordinate having values between 0 and 10. This corresponds to simply picking 10 random integers between 0 and 10 and thinking about them as a coordinate. So, one possibility is (7, 0, 9, 10, 7, 8, 8, 4, 7, 1). Now, if we create five such random 10D points and visualize them as Chernoff faces, we get the images in Figure 17.2.

Let's move into 13D with higher dimensional sports data. Dr. Steve C. Wang of Swarthmore College used Chernoff faces to graph managers in MLB[55]. He was interested in comparing their managerial styles, in particular who were similar. We'll go back and look at the 2008 season. Managers were compared on such statistics as the rates of bunting, stealing, pinch-hitting and number of different line-ups. Keep in mind that while the face is plotted in two dimensions, it encapsulates more than two data points. Wang used 13 (Figure 17.3).

St. Louis Cardinals manager Tony La Russa used 140 different batting orders (158 if you include pitchers) in the season's 162 games. So, his Chernoff face has an elongated head with wider eyes. In contrast, the Philadelphia

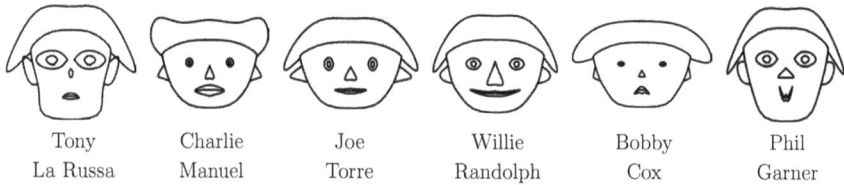

| Tony
La Russa | Charlie
Manuel | Joe
Torre | Willie
Randolph | Bobby
Cox | Phil
Garner |

Figure 17.3: Chernoff faces of managerial styles in MLB enable us to plot in 13 dimensions.

Phillies' Charlie Manuel used only 77 batting orders (117 if you include pitchers). Manuel's Chernoff face is squattier with beady eyes.

So, who looks most alike? Which Chernoff faces could be siblings? Look at the face for Mets' manager Willie Randolph and Yankees' manager Joe Torre. They look a lot alike. Looking at their history, Torre and Randolph were both with the Yankees where Torre was Randolph's managerial mentor. On the other hand, Bobby Cox and Phil Garner look distinctly different.

Chernoff faces are interesting but must be used with care. We have a tendency to perceive some features more than others. For instance, eye size and eyebrow-slant tend to carry more weight in how we perceive a face. Said another way, which stats correspond to which facial features can affect what we notice. Yet, this graphical technique does underscore how we can explore higher dimensional data graphically.

17.2. Seeing 3D Two Dimensionally

The 2D world exists in a plane. Let's take a page of this book as our plane. You and I exist in the 3D world so let's imagine if objects we hold can pass through a page of this book. Suppose you were holding a ball and you could pass it through the page, as seen in Figure 17.4(a). A circle intersects the page. If we move the plane vertically by 1 unit and keep track of the circle formed by the boundary of the sphere intersecting the plane, we form the graph in Figure 17.4(b).

Note how the circles are not equidistant from each other. Some are closer than others. Keep in mind that, we moved vertically by 1 unit each time. When the circles are closer to each other, we are at a point on the sphere that is vertically steeper. If we were hiking, this would be a harder region to climb. The same idea holds for topographic maps. In Figure 17.5(a), we see a detail of such a map of Yosemite National Park in California. Look

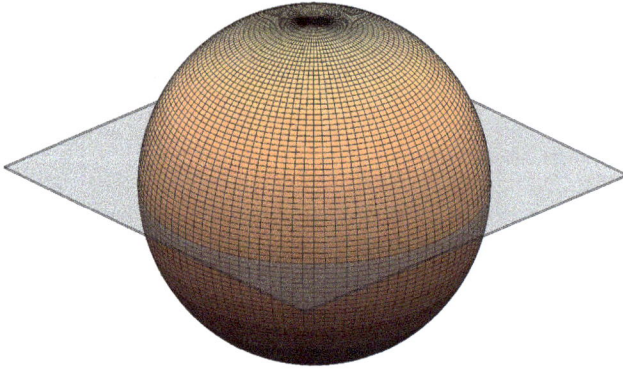

(a)

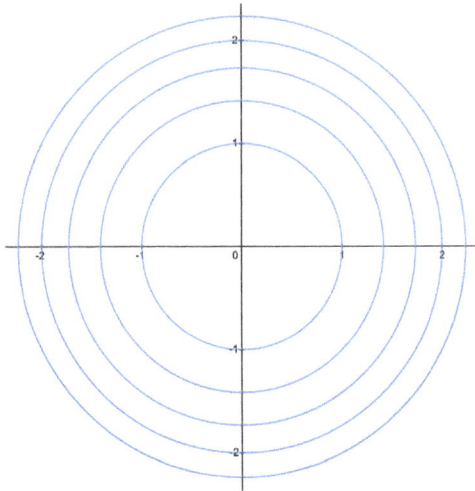

(b)

Figure 17.4: A sphere intersecting a plane and the circles that will intersect the plane at different heights of the circle.

for the words "Half Dome" and note how close the lines are. This is a very steep region. Compare the distance of these lines to the green regions to the left of "Half Dome". That green region is part of the Yosemite Valley where visitors camp and drive their cars.

Let's return to Half Dome. How steep is that region? The lines are close but how steep? Just take a look at Figure 17.5(b). It's very, very steep. Note how the shape of Half Dome has similarities to a sphere.

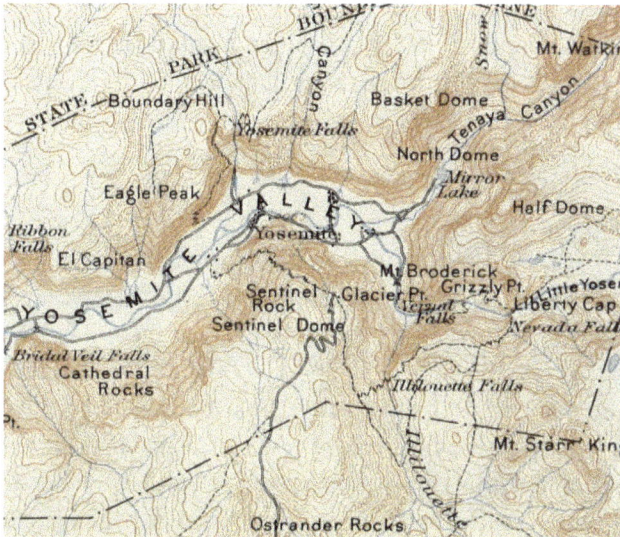

(a)

(b)

Figure 17.5: A detail of a 1905 topographic map of Yosemite National Park by the United States Department of the Interior Geological Survey (a) and Half Dome (b) as seen in the park.

How about a cube? What curve results when it intersects a plane? Clearly, we can make a square. However, that isn't the only shape. If the plane intersects with a corner of the cube, then we get a point. So, the angle of the plane to the cube makes a difference. In fact, you can find planes that will intersect a cube to form a triangle, rectangle, and hexagon. Can you figure out how? Figure 17.6 (a) contains a blank cube for you to explore and discover yourself. The answers are at the end of the chapter.

17.3. Seeing a 4D Cube

Let's take another look at the cube. In Figure 17.6(a), a 3D cube is projected into 2D. What would a 4D cube look like projected into 3D? We can't represent that on a 2D page of paper, so in Figure 17.6(b), we see a 4D cube, called a hypercube, projected down into 3D and then that 3D projection again projected into 2D so it can be printed on the page.

There are other ways to represent a 3D cube in 2D or a 4D hypercube in 2D. For example, we can look at a net of a polyhedron, which is an arrangement of polygons in the plane which can be folded (along their edges) to become the faces of the polyhedron. For example, Figure 17.7 are three ways we could represent a cube. Think about how to fold each of these into a cube along the edges.

This helps us to think in a different way about a 4D hypercube. In Figure 17.8(a), we see a net of a hypercube. This also appeared in Salvador Dali's 1954 oil painting *Crucifixion* (*Corpus Hypercubicus*) as seen in Figure 17.8(b).

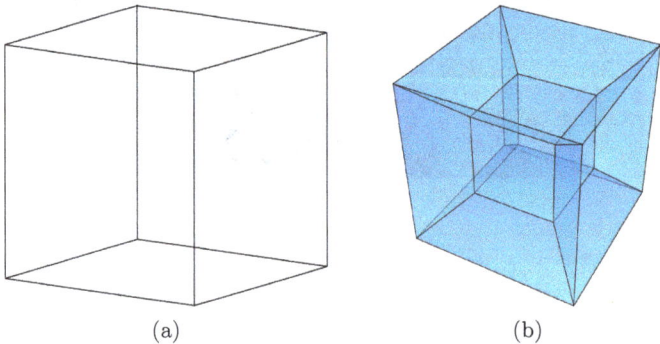

(a)   (b)

Figure 17.6:   Use the blank cube in (a) to find planes that will intersect to form a triangle, square, rectangle, and hexagon. In (a), we also see a 3D cube projected into 2D, which enables us to see it on the page. In (b), we see a 4D cube projected into 3D and then to 2D.

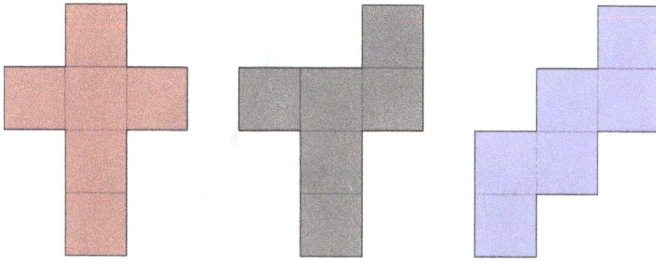

Figure 17.7: Each of these images can be folded into a cube. Images by R. A. Nonenmacher.

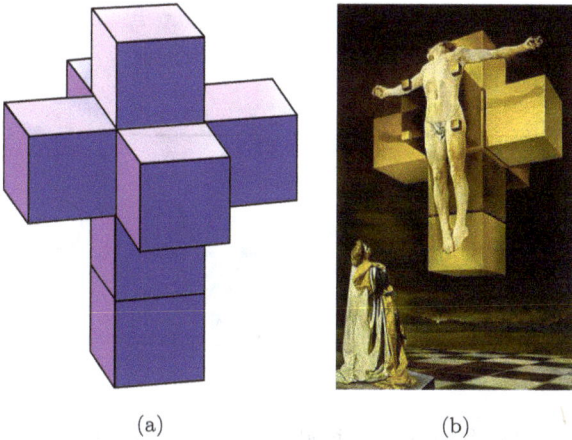

(a)   (b)

Figure 17.8: A net of a 4D hypercube (a) (image by Robert Webb) and Salvador Dali's *Crucifixion* (*Corpus Hypercubicus*) (b).

17.4. Understanding a 4D World

Dali's painting contains a representation of a 4D object in a 3D world. Let's think more about this type of idea. For this, let's consider my self-portrait in Figure 17.9.

What if I lived in the 2D world of this page? If I looked like this cartoon, I would have very limited vision since my eyes couldn't see beyond the edges of my face. Any line in two dimensions is analogous to a wall in our 3D world.

Figure 17.9:   My 2D self-portrait.

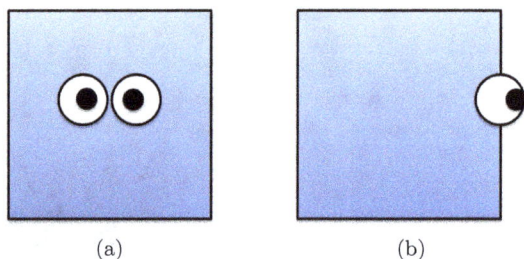

(a)   (b)

Figure 17.10:   Two 2D blockheads, is one more realistic than another?

Let's get even simpler by using the drawing in Figure 17.10(a). The pupils of this 2D blockhead could only see as far as the edges that outline its eyes. What cartoon might lead to a happier existence in such a 2D world? How about the picture in Figure 17.10(b)? Now, our square friend can see the world and appreciate the words on the page.

What if our blue cartoon still lived in two dimensions but his world, rather than being flat, was curved like the surface of a sphere? He could, much like those in the time of Christopher Columbus, think that his world was flat. In such a 2D universe, the world could seem flat, when it actually had curvature in the third dimension where you and I live. We 3D human beings only fully appreciate the curvature of our world when we leave the seemingly 2D surface of our planet and view it from outer space — something that the 2D blockhead cannot do.

In a similar way, could our 3D universe be curved in a fourth dimension? Could this even be detected in our third dimension? Let's think about living on a sphere and the angles of a triangle. If you draw a triangle on a piece of paper, the sum of its angles form a straight line. However, if you connect three points on a sphere, the angles sum to greater than 180 degrees. This could be an indication that, even if it seemed like one lived in 2D, one's world is curved.

This has been a question about our universe. Einstein's theory of general relativity states that space itself can be curved by mass. As a result, the density of the universe determines its shape. The exact shape of our universe is still a matter of debate. Yet, experimental data, repeated by independent sources, have found that the universe is flat with only a 0.4% margin of error.

17.5. Close but in a Higher Dimension

What relationship could someone in the 4th dimension have with our 3D universe? We can consider the question by looking for analogs in lower dimensions.

Let's start in one dimension, which is a line. Suppose a small circle, which exists in 2D, is close but not intersecting the line as seen in Figure 17.11 on the left. The circle is close in two dimensions but not perceivable within one dimension. If the circle intersects the line, then it becomes visible in 1D, but only a portion of it. In this way, the circle exists in 1D while still existing in 2D.

Let's move up a dimension and consider someone in 3D interacting with a 2D universe. Think about your place within the 2D world of this page of paper. You could put your nose centimeters from the page containing Figure 17.10(b). As long as you don't punch through the page in front of the blockhead's eye, our flat friend wouldn't know you were there. You are close but invisible as you exist in a higher dimension.

Figure 17.11: A 2D circle can be close but not seen in a 1D world as seen in the left or intersecting the world as seen on the right.

We can also reflect again on the sphere intersecting the plane in Figure 17.4(a). The sphere is always a sphere in 3D whether or not it intersects a 2D plane. However, when intersecting the plane, the sphere would look like a circle like those in Figure 17.4(b). Somewhat mystically, if the sphere no longer intersected the plane, it would disappear but could still be closeby. Rather than a sphere, imagine having a cube that became its unraveled self as in Figure 17.7. The next moment, it could reassemble itself back into 3D and only have a portion of itself intersecting the square. Akin to Dali's painting in Figure 17.8(b), a 3D cube would be entirely contained in 2D. Dali's painting had Christ upon a 4D cube that's contained within 3D.

This chapter can raise a variety of questions. Could the shape of our universe be curving in some higher dimension? Could someone in the fourth dimension be present, like you could be to our blockhead, without being visible in our 3D world? What would a 4D sphere look like were it to intersect our world? In part, this depends on what part of the 4D sphere intersects the 3D world. We saw this when we asked, what curve results when a cube intersects a plane. We mentioned that it can be a point, a triangle, a square, a rectangle, and a hexagon. Did you figure out why? The answer follows.

Spoiler Alert

When a cube intersects a plane, it can result in a triangle, a square, a rectangle, and a hexagon. Below you can see how such shapes can be formed.

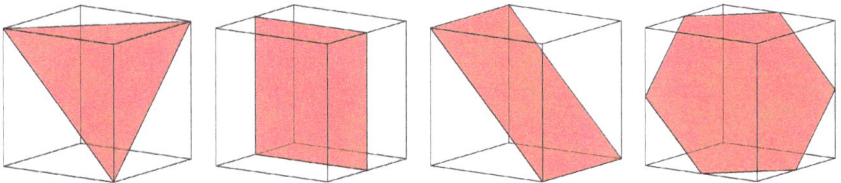

Chapter 18

Tied in Knots

Get some string, preferably slippery, like a lanyard and tangle it up — either haphazardly or methodically, your choice. Then, tape the ends together. Untangle it as much as you can. What do you get? Can you untangle it and get a loop as in Figure 18.1(b)? If not, could you untangle it some other way to get the loop? My tangle is seen in Figure 18.1(a). I was able to untangle it to get the loop in Figure 18.1(b). But did I break my string, untangle it, and then reconnect the ends? Can you even know? A branch of mathematics called topology considers these types of questions. The loop in Figure 18.1(b) is also called the trivial knot.

Math often looks at equality. For example, $x^2 + x = 0$ implies $x = -1$ or $x = 0$. In order for $x^2 + x$ to equal 0, we must have x equalling -1 or 0. Topology studies the equivalence of objects. We do this with objects. For instance, in Figure 18.2, we would categorize (a) and (b) as chairs but not (c). Note that, we can sit on any three of them and both (a) and (c) have four legs. Still, we see (a) and (b) as chairs. In this sense, we are able to classify when objects are the same and when they are not. Important to this chapter is the idea that (a) and (b) are not exactly the same as they differ, but we can still group them by their similarities.

In this chapter, we'll do a similar thing and talk about equivalence. Objects will be grouped by mathematical similarities of interest. We do this with numbers when we talk about numbers as being odd, even, prime, positive, negative, rational or irrational. We'll consider the equivalence of shapes rather than numbers in the remainder of this chapter. For example, if you have a tangled mess like you made at the beginning of the chapter and can untangle it into the loop, without cutting the rope or disconnecting

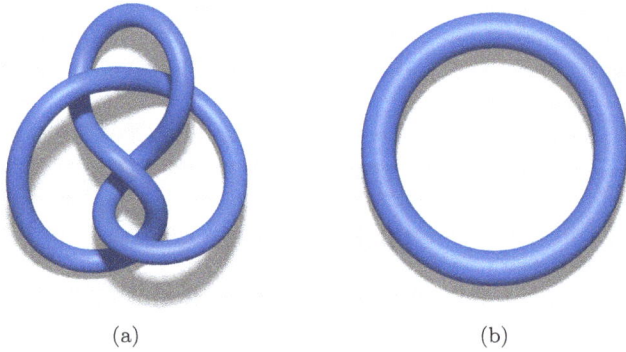

(a)   (b)

Figure 18.1:   Can you get from the tangled mess in (a) to the loop in (b) without breaking the string?

SAME   SAME   NOT SAME

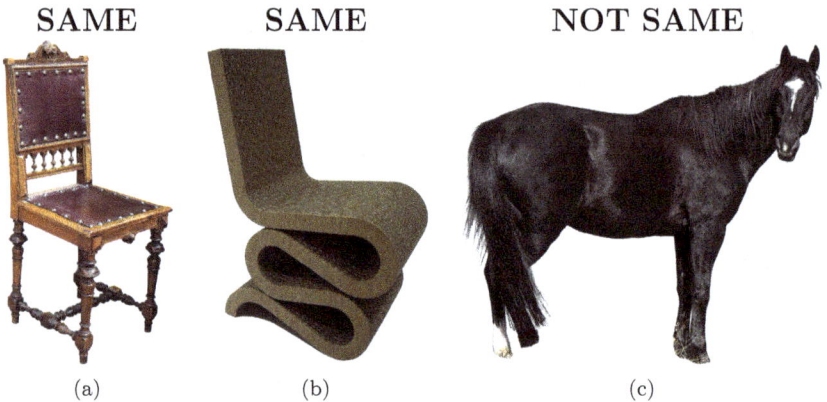

(a)   (b)   (c)

Figure 18.2:   We often group objects by their similarities like (a) and (b).

the taped ends, then we consider that tangled loop as equivalent to the loop. That loop is called a trivial knot or an unknot. It turns out that my loop in Figure 18.1(a) is not equivalent to the loop in Figure 18.1(b). Topology is a field that would enable you to know this without touching the string or fiddling your way through untangling.

Topology is sometimes called rubber sheet geometry. To get a sense of this, let's play another game with a piece of rope. This time, the rope can shrink or stretch as needed. The rules are simple. We'll have initial and target shapes. Can we move from the initial to the target shape without detaching or attaching the ends of the rope? If it is possible to do so, the shapes are equivalent, else they are not.

18.1. Bendable Alphabet

Suppose we put our string in the shape of the letter L as seen in Figure 18.3? If we take each of the other pictured letters as our target shape, which of them are equivalent to the letter L?

If you think of the string as a caterpillar, it need only straighten itself to move from the L to the I. It need only curve its lower half and curve its whole body to form the J and C. The O presents an issue as there is simply no way for a rope with 2 exposed endpoints to form a shape with no endpoints. So, L is equivalent to all the letters except the O in Figure 18.3.

Let's try again. Suppose you form a single cord with 3 ends into the letter T as seen in Figure 18.4. Which of the other letters are mathematically equivalent?

It doesn't take much to get from the T to the Y. It can take some contortions of the mind to see how to get from the T to the E, although some find it easier to imagine going from the Y to the E. Seeing that you cannot get from the T, E or Y to the K can be difficult as you, of course, must

L I O
J C

Figure 18.3: An alphabet game.

T E
K Y

Figure 18.4: Another alphabet game.

imagine what is impossible. One helpful insight is to note that the K has four endpoints. The other letters have 3. You cannot get from three to four endpoints without detaching the string or cord at some point. So, the letter K is not equivalent to the other letters.

18.2. Hand-to-Hand Mathematics

You can also try this idea with a group of friends. Each person grabs two other person's hands. If you try this with two friends and no one takes their own hand or more than one hand of another person, then your group will form a configuration like that in Figure 18.5(a). We can also form a similar configuration with a colored string as seen in Figure 18.5 (b). Figure 18.5(a) and 18.5(b) depict what is called the trefoil knot. The string is colored to help make associations with the people in (a). You can't unknot the trefoil knot to form the trivial knot in Figure 18.1(b).

Suppose you have a large group of friends and each person grabs two other people's hands. Here is a question: is the group connected in one continuous loop or did you form two or more disconnected loops? Why didn't this question arise for a group of three people? By stating that you can't take your own hand or both of another person's hands, we excluded the ability to make disconnected loops. If you remove these restrictions, a group of three people can make configurations like those in Figure 18.6.

Did you figure out how to test if your tangled group made disconnected loops? One person can start by squeezing the hand of the person holding

(a)   (b)

Figure 18.5:   You can make knots with a group of people or with a string. Both (a) and (b) are in a configuration called the trefoil knot.

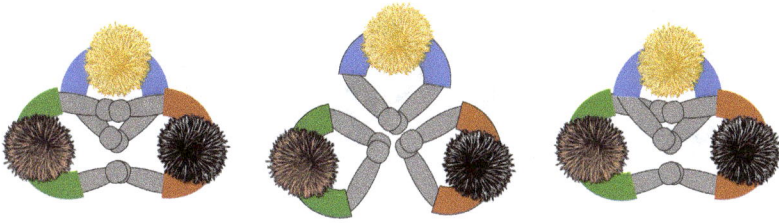

Figure 18.6:   Some configurations of three people grabbing random hands in the group.

their right hand. The person who just had a hand squeezed, squeezes with their other hand. This continues until the first person's left hand is squeezed. If everyone's hands were squeezed, the group made a continuous loop. Else, there are disconnected loops. To find out how many loops, someone whose hand wasn't squeezed can initiate the same process to find out if all the remaining people are in the other loop or several loops. If your group formed one continuous loop, then your group formed a knot.

Here is another question for the group: if the group formed a knot, can the group unknot into the trivial knot, which is similar to our question with your string in Figure 18.1? Be fore warned. This takes some work from the group and can look a lot like the game Twister. If the group has formed disconnected loops rather than one continuous loop, can the disconnected loops be pulled apart without any releasing hands? Note that, figuring out if a knot can become the trivial knot is a difficult question for complicated knots.

18.3. Flexible Thinking in 3D

Let's move from knots to shapes. Remember that we are considering rubber sheet geometry where things are extremely flexible. They stretch and bend as much as needed. We can't disconnect or connect anything — just stretch. Here are the puzzles. Can you get from the shape in Figure 18.7(a) to the shape in Figure 18.7(b) with only stretching? The second puzzle is the same question but a different set of shapes. Can you get from the shape in Figure 18.7(c) to the shape in Figure 18.7 (d) with only stretching? A hint on

(a)   (b)   (c)   (d)

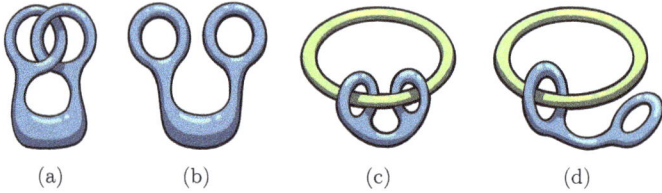

Figure 18.7:   Can you get from the shape in (a) to the shape in (b) with only stretching the shape and not disconnecting or connecting components? Can you get from the shape in (c) to (d) with only stretching? Illustrations by Ansley Earle.

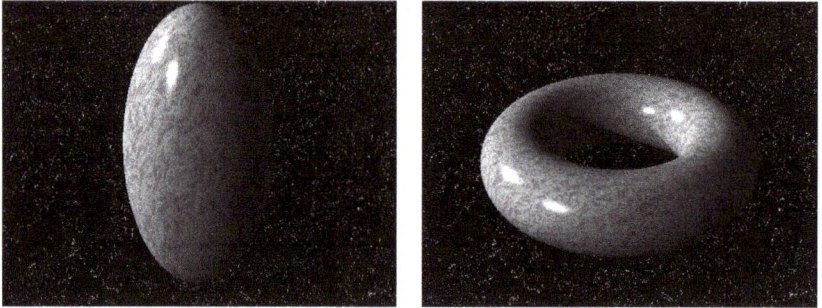

Figure 18.8:   How would you know if you landed on an egg-shaped or donut-shaped planet?

both questions is given in the footnote when you are ready for it.[a] The actual answer is at the very end of the chapter. Work on it awhile and see what you think!

Now, let's consider the mathematical equivalence of a donut and an egg. Suppose some day in the future, an astronaut is traveling through space and crashes on a distant planet. She has only a compass (which we will assume points north at all times) and the very large amount of debris from the fallen ship. By leaving a sufficiently long trail of debris (assuming nothing on the planet moves the trail), how could the astronaut determine whether the planet is in the shape of an egg or a donut as seen in Figure 18.8? Note that, the planet is covered in clouds so she would never be able to look up and see the planet above her, which would be possible on some points of the donut-shaped planet on a clear day.

---

[a] You can get from Figures 18.7(a) to 18.7(b) and Figures 18.7(c) to 18.7(d) with stretching! Both challenges are possible.

The hole in the donut-shaped planet is a key. You're stuck on the planet so you have a lot of time to create debris trails. You'll need to leave more than one trail of debris around the planet. If you are on the egg-shaped planet, you'll loop around the planet leaving a trail of debris. When you return to the staring spot, change your angle. Every other trail you create around the planet will have two crossings with every other trail you've created. If you are on the donut-shaped planet, it is possible for trails to only cross once. To see this, suppose one trail can go around the outer edge of the donut. When you return to your starting spot, turn 90 degrees and you'll go through the hole of the donut-shaped planet. You'll intersect the other trail but only when you return to the starting spot. This is an important difference in these shapes and is a way to see they are not equivalent.

18.4. Flexing Data

How does this connect to data analysis? Suppose we have the points in Figure 18.9(a) versus the points in Figure 18.9(b). How do they differ? A field called Topological Data Analysis, or TDA, can aid us. Topology helps uncover the geometric structure of data. TDA can help find essentially a cloud around a set of points. The points in Figure 18.9(a) are contained largely within an elliptical cloud. In contrast, the points in Figure 18.9(b) can exist largely within a cloud resembling the letter L. We see the clouds in Figures 18.9(c) and 18.9(d).

We've seen from our earlier work with the alphabet that an ellipse and the letter L have different properties and can be seen as different shapes. Topology helps find such structural differences.

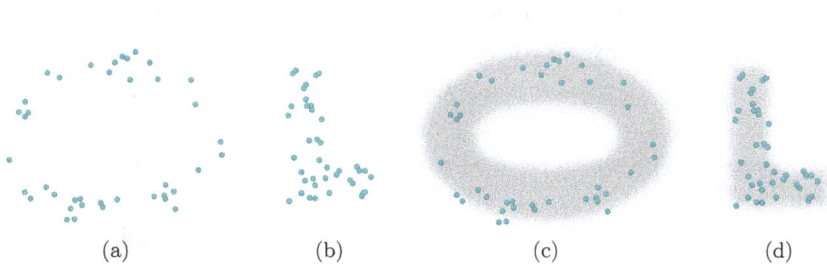

(a)   (b)   (c)   (d)

Figure 18.9: Topology can help find geometric structure differences in data.

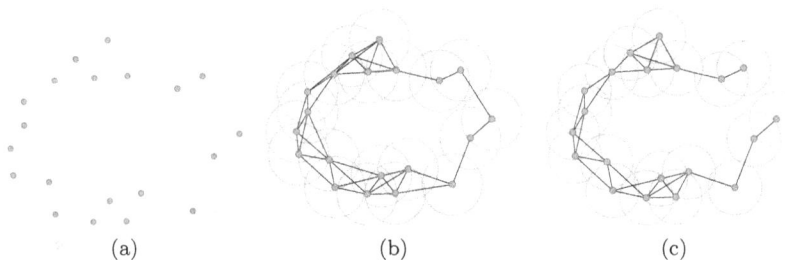

(a)   (b)   (c)

Figure 18.10:   Topological Data Analysis can turn data points into shapes.

We need our points, like that in Figure 18.10(a). How do we get from a set of points to a shape like the letter L in Figure 18.9(d)? Here is a procedure:

1. Pick a distance d.
2. Draw a circle with radius d around every point.
3. If two circles intersect, draw a line between the two points at the centers of the circles.

If we try this with the points in Figure 18.10(a), we get the circles in Figure 18.10(b). The intersection of the circles looks like an ellipse. We did it. For our particular d, we did. Let's reduce d by 15%. Now, we have the shape in Figure 18.10(c). Suddenly, our shape is the letter C, which is topologically equivalent to the letter L, as we saw earlier. While we have an algorithm, the shape isn't unique. For different distance values, the geometric structure of the underlying shape can change. Still, this algorithm can be used to understand differences in data and it doesn't necessarily need to be in 2D.

TDA can be applied to higher dimensions. Let's turn to sports and work in the seventh dimension. At the 2012 MIT Sloan Sports Analytics Conference, Muthu Alagappan, who was a Stanford University senior at the time, showed his work in TDA as an intern at Ayasdi, a data visualization company. Alagappan looked at seven stats for all 452 NBA players from the 2010–2011 season. The work explored the seventh dimension from seven statistical categories: points, rebounds, assists, steals, blocked shots, turnovers, and fouls. The stats were normalized on a per-minute basis.

In the end, Alagappan's work defined 13 positions in basketball: Offensive Ball-Handler, Defensive Ball-Handler, Combo Ball-Handler, Shooting Ball-Handler, Role-Playing Ball-Handler, Three-Point Rebounder, Scoring Rebounder, Paint Protector, Scoring Paint Protector, NBA 1st-Team, NBA 2nd-Team, Role Player, and One-of-a-Kind. Naturally, the positions can be described differently. For example, the Offensive Ball-Handler handles the ball and specializes in points, free throws, and shots attempted, but is below average in steals and blocks. Tony Parker was an example of this player. In contrast, a Defensive Ball-Handler is a defense-minded player who handles the ball and specializes in assists and steals, but is only average in terms of shooting. An example in the dataset was Kyle Lowry. Finally, Shooting Ball-Handler is above average in field goal attempts and points. Stephen Curry and Manu Ginobili were examples from the data. Rondo and Battier were both Role Players in Alagappan's work. While one can debate if NBA 1st-Team or One-of-a-Kind are positions on a team, the work classifies players. The work received attention appearing in *Wired* and *Slate* [5,34].

Note that, when we work in higher dimensions, this is akin to comparing a higher dimensional square to a higher dimensional torus. Topology enables us to compare and analyze such shapes after our work transforms the data into shapes.

Spoiler alert

Let's see the answers to the challenges in Figure 18.7. First, both challenges are possible. You can get from Figures 18.7(a) to 18.7(b) and Figure 18.7(c) to 18.7(d) with stretching! Now that established, the question is how! The following illustrations are by Ansley Earle.

Challenge 1: Let's see first how to get from Figures 18.7(a) to 18.7(b).

Challenge 2: Let's see first how to get from Figures 18.7(c) to 18.7(d).

Do you find these answers unbelievable? Remember that, topology assumes that the shapes can stretch as far as possible. Sometimes, that is farther than our intuition allows objects to stretch.

Chapter 19

More Sporting Lessons

We've covered a number of topics since our first set of lessons in Chapter 10. As we move to the end of this book, let's return to another set of lessons from sports analytics and mathematics.

19.1. Unknown until Known

Mathematics can involve a lot of effort. Sometimes, the mathematics necessary to accomplish something is difficult and sometimes it is easy. However, until it is known what to do, it is simply unknown. Let's see such an example from a million dollar challenge!

Suppose we are sitting on a couch flipping through television channels and stop to watch a clip of the 1992 film *My Cousin Vinny*. I comment, "I love this film!" Would you mention other films that you could recommend that you expect I would like, too? Such a knack for recommending could have earned you a million dollars from Netflix about a decade ago.

To enter the millionaires' club, you needed to do much more than recommend a film to me on the couch. In a sense, you had thousands and thousands of people sitting on the couch telling you the films they liked and how much they liked them. Rather than a couch, you had access to Netflix ratings data. To win the case, you needed to do a better job at predicting than Netflix's recommendation system, called CinematchSM, which predicted whether someone would enjoy a movie based on how much they liked or disliked other movies. Specifically, if your predictions are at least 10% better than

CinematchSM for this training set, you won the Netflix Prize worth a million dollars! Oh, there was a caveat — you had to share the results with Netflix and "describe to the world how you did it and why it works".

The competition began on October 2, 2006. Netflix provided a training dataset of 100,480,507 ratings that 480,189 users gave to 17,770 movies. Each training rating was a quadruplet of the form (user, movie, date of grade, grade). The user and movie fields were integer IDs, while grades were from 1 to 5 (integral) stars.

By October 8, a team called WXYZConsulting had already beaten Cinematch's results. By October 15, there were three teams who had beaten Cinematch. Within two weeks, Netflix, which was a DVD rental company at the time, had received 169 submissions. By June 2007, over 20,000 teams had registered for the competition from over 150 countries.

The quick improvement in those opening weeks made Netflix's task look simple. But fortunes changed as the rate of improvement slowed. Month after month, the same three or four teams topped the leaderboard, ticking off decimal by decimal improvement. After a year, BellKor, a research group from AT&T, was in first place beating Cinematch by 8.43%. At the same time, murmurs spread about the potential impossibility of 10% improvement.

Suddenly, in November 2007, a new entrant, "Just a guy in a garage", appeared with a score that not only popped into the top 10 but performed at a level that had taken BellKor seven months to achieve. By mid-January, there were just five teams, out of 25,000 entrants, ahead of the garage guy entrant. Adding to the mystery, no one knew the identity of the entrant or what statistical sorcery were sprinkled into his methods.

It turns out that the entrant, Gavin Potter, didn't work out of a garage but instead a back bedroom of his London home. The 48-year-old Englishman was a retired management consultant with an undergraduate degree in psychology and a master's in operations research. As he developed his predictive algorithm, Potter turned to his oldest daughter, who would break from her high school homework to serve as her father's math consultant and solve various tasks requiring Calculus. Potter's improvement came not through advanced stochastic processes but an area yet untapped in the competition: human psychology.

Remember, the ratings were made by real people. Potter's work focused on predictive patterns stemming from our humanity. For example, the Netflix dataset contained eight years of ratings. Perspectives often change over time, which can necessitate weighting recent ratings more heavily than older ones.

A key to Potter's success lay in behavioral economics, a field pioneered by Amos Tversky and Nobel Prize winner Daniel Kahneman. In particular, consider the anchoring effect, which can blur any rating system. Suppose you watch three straight movies that you rate with four stars. If you then see *The Avengers* and deem it slightly better than your previous three four-star rated films, then you are likely to rate this Oscar nominated film with five stars. But, if you started that same week with a one-star film, then *The Avengers* might only get a 4- or even 3-star rating. Potter found that the Netflix data contained this inertia in rating movies and integrated it into his methods.

Why didn't the other competitors observe this effect in the dataset? Simply put. They didn't know to. Potter's training in psychology and knowledge of behavioral economics focused his attention on something new to the community. It's easy to imagine many of those top teams thumping their hands on their foreheads upon learning Potter's methods and exclaiming, "Of course! That makes sense."

I've heard many data scientists state "we often don't know what we don't know until we do." Things remain unknown until they are known. It's hard to know if the Netflix Prize would have been won without the insights of behavioral economics. Those murmurs on the competition's intractable request may have, in time, been considered truth.

In the end, the march toward 10% was more of a long distance run than a sprint. Finally, in September 2009, a conglomerate team that named itself BellKor's Pragmatic Chaos won the million dollars, beating Netflix's algorithm for predicting ratings by 10.06%. While there are many stories and lessons in data science from the competition, we learn an important one from a guy in a garage, or back bedroom. Be careful not to think you know all that's needed to succeed or optimize a solution in data science. A huge improvement may be just a new insight away. Uncovering some bit of information could be a breakthrough that transforms your work and springboards your productivity.

19.2. Sick of Data

When winter approaches, do you worry about catching the flu? At one time, the Center of Disease Control (CDC) was the hub of information on possible outbreaks of influenza. Their warnings came from doctors reporting new flu cases. When the CDC's analysis indicated an emerging pandemic, the news was outdated by a week or two. Today, we can access nearly real-time information by, as we often say, googling it.

Google's work in predicting flu outbreaks began when the company examined the 50 million most common search terms that Americans type. They also looked at CDC data between 2003 and 2008. They searched for correlations between frequencies in search queries and the spread of flu over time and space. A combination of 45 search terms led to a strong predictive ability. Like the CDC, Google could pinpoint locations of flu outbreaks. Google's results were, however, considerably faster since their model could begin to see a pattern before some search engine users walked into a doctor's office or got the results of a mouth swab. This epidemiology work by Google was published in *Nature* magazine in 2009, claiming they could "nowcast" the flu. Soon afterwards, the country entered an H1N1 crisis. Google's results were more accurate than the CDC and occurred essentially in real time.

Google offered the tool online, which was extended to more than two dozen other countries. The site, Google Flu Trends, estimated current outbreak levels in the US, and could be refined to examine data at the state and some city levels.

Google's ability to essentially don an electronic lab coat and diagnose H1N1 outbreak levels in 2009 garnered well-deserved attention. The 2012–2013 flu season sent Google down a different path, one of sharply overestimating the amount of flu. Google Flu Trends warned that nearly 11% of the population were infected, when follow-up information from the CDC found no more than 6% had caught the flu. What went wrong? Media reports about influenza, including a state of emergency declared in New York, led to more searches that year by people who weren't sick.

Google had already anticipated how media coverage of the flu could lead to spikes in searches by healthy people for three to seven days. But this time, the flu season actually was worse than the previous year, and the searches actually continued at a higher level throughout the season. More media coverage and more flu meant that even healthy people did much more searching for the term "flu". The assumptions that successfully aided during the 2009 H1N1 crisis lay at the core of its failure in 2013.

It's insightful to focus on Google's response. Seeing the challenges of their epidemiology predictions in 2013, Google adjusted their algorithm. They first adjusted for spikes in searches after media coverage. They also refined how their linear regression throws out extreme values. With these changes, their revised method once again gave an accurate estimate of flu, within 1% of what CDC data reported.

Our lesson is less about Google's initial success and much more about its response to the method's struggle. Algorithms, like individuals, must be examined for their health and strength. An annual doctor's exam can uncover health concerns before they manifest into chronic challenges. Even so, we get sick and must recover. To maintain strength, methods must be examined — given a check-up for algorithmic health. Nonetheless, algorithms can fail. Today's cutting-edge technique may soon become an older version as algorithms often evolve. We must check in on algorithms, else we simply assume their health. There is a danger if we simply wait for algorithms to become dramatically failing in their analytical health. It isn't always clear when the flu season will come, but we can take steps to be ready and healthy before it arrives.

19.3. Searching the Signal

How do we know we can recommend a movie or that the recommendation means anything? How do we know that a sports analytic can give insight? Much of this has to do with searching for signals that enlighten us on information. Let's look more closely at searching for a signal in analytics.

The Titanic set sail on April 10, 1912 with 2,224 passengers and crew traveling from Southampton, England headed to New York City. The Titanic was the largest ship in the world and was considered unsinkable. Four days into the journey, the ship suffered a glancing blow to an iceberg and under four hours later, the Titanic broke apart and sank, with well over one thousand people still aboard.

It took over 70 years to find where the passenger liner lay at rest at the ocean's floor. While many searched for the ship, it remained hidden in the darkness of the sea until Robert Ballard, an ocean explorer, realized that a vessel's wreckage was most likely carried by the undersea currents, leaving a tail of debris. In 1985, Ballard found the debris trail, which led directly to the Titanic. The discovery became possible when Ballard found the right trail.

Similar quests for discovery happen in our modern sea of data. The signal that can serve to guide our exploration is often finding the right datasets to combine. As a digital example of this type of exploration, let's consider my analysis of the *Game of Thrones* television series.

Since its first season, *Game of Thrones* seemed to guarantee two things — winter was coming and, second, anyone might die in the show. But death was difficult to predict leaving fans often mulling over the options between episodes and seasons.

Patterns exist in many places, even unexpectedly where we might perceive total randomness. In my role as Chief Academic Officer at the predictive analytics software company, Tresata, I decided to collaborate with an intern and examine data related to the show. We looked at the data prior to Season 7, which would air that July. To begin our work, what data should we examine?

Data can become an asset when multiple sources are combined. Sometimes, $1 + 1$ does equal 2 as you aggregate two datasets and simply get back what each originally offered. Other times, it is as if $1 + 1 = 500$, with aggregated datasets offering a rich bed of insight and discovery. The trick is what data to combine. This is usually not obvious and can even be unexpected.

In the case of our analysis of the *Game of Thrones*, we began by collecting various datasets. We knew we could slice and dice the data, which enabled us to find connections that might otherwise be hidden. In the end, insight came from simply combining two datasets, among the many we collected. The first was a site (http://deathtimeline.com/) listing the chronological order of deaths in *Game of Thrones* from seasons 1 to 6. This site detailed the time of deaths, who died, and whether a fallen character was classified as minor, recurring, or major. The key came when this timeline was combined with information on the writers for episodes, taken (and verified) from Wikipedia (https://en.wikipedia.org/wiki/List_of_Game_of_Thrones_episodes).

What we found was that in seasons 1–6, except for a character named Joffrey, every major character in the series died in a script written by David

Benioff and D. B. Weiss. We found other insights, too, such as when the rate of deaths increase, when to expect major battles, and when in an episode to expect deaths.

A few weeks later, I watched a new episode in season 7 of *Game of Thrones* with my wife. As the opening credits came to a close before the first scene of that night's episode, my wife heard me sigh with relief. She asked, "Did you just figure something out?" I responded, "Yes, I just got a signal about tonight's episode. Would you like to know?" She reflected for a moment and then agreed.

That evening's episode was not written by David Benioff and D. B. Weiss. So, I shared with my wife that I did not expect a major character to die that evening. While that night's script moved us through the customary rollercoaster of drama, there were no deaths of major characters no matter how precariously close they seemed to step into death's doorstep. As we turned off the television, my wife summarized my predictions with her comment, "That was some insightful data!" The insight came from combining the "right" datasets. Until then, such observations would have remained obscured in apparent randomness.

If you believe there may be patterns in a phenomenon even if it appears to be largely random, consider exploring data. Keep in mind, however, that a key step can be finding the right data, else such insights can remain hidden. In analytics, there is the signal and the noise. Sometimes, the signal can be found and other times the phenomenon is noisy randomness. It can be difficult to predict which is which and it can be a game of insights where you have to wait, when winter comes, for it to pass as you spring into insight.

19.4. Analytics Antenna

During my childhood, my family had an old TV in the basement. It's reception came via an antenna that sat on the top of the rickety set. To watch a program, I'd fiddle with the rods hoping to see images emerge through the fuzz. Analytics is often the same. Life has its serendipity. Such events aren't predictable but are generally inherent. Within the noise of randomness, analytics can often identify and even quantify a trend. Yet, seeing the signal isn't always easy.

Sports analytics is a subfield of data analytics and encounters many of the issues that arise in financial, retail, or health care analysis. From a last second shot that bounces around a rim to end a game in victory or defeat to balls that are tipped on the gridiron and result in interceptions or miraculous receptions, sports has the random elements that can obscure the underlying trends of skill and talent.

With my family's television set, I easily saw when broadcast images emerged into the picture. In data, seeing a trend emerge isn't always easy. In fact, even when a pattern appears, one must ensure its generalized presence.

To give more context, let's turn to my research. I've worked with ESPN on an analytic to identify unpredictable teams. For each season, we analyzed over 5,000 games involving approximately 350 Division 1 NCAA men's basketball teams. I wrote a computer program to analyze the data with our new algorithm. The result was a list of the teams classified as unpredictable and predictable.

One rarely knows how analytics will unfold until results emerge. And results are rarely perfect. Randomness plays a role. A high degree of accuracy often indicates promising, new results.

For this work, I analyzed seasons spanning over a decade. Sometimes, early results led to only a moment of laughter and then reflection regarding additional steps that might improve the work. Then, there are moments, like this instance, where the results make me pause and look again. When I looked at the first list, I saw teams classified, as I would hope. I looked at more seasons. Every year, I saw teams classified correctly, but not all of them. I didn't expect perfection as one analytic rarely captures every variation of a phenomenon.

As I looked at the results, I was admittedly stunned. I paused, took a short walk, and then returned again and looked at the results. I did appear to have a metric that classified teams with attributes of interest. One of my collaborators, an alumnus of Davidson College with whom I've done a lot of research both during her time at the college and now as an alumnus, asked about the work. I showed her the lists.

She turned and said, "Wow! We might have something. And, we may not. We need to automate the process and see if a computer can find what we are seeing." I smiled as that was a speech I often gave. "Why?" I asked.

"Because we might only be finding what we want to find and overlooking ways that this result isn't as easy to see as it seems." She was right. Did the metric work? In time it did and was covered by *ESPN Magazine*.

Let's pause and see important parts of this process. We had a new idea on data that could not be easily analyzed by hand. Further, we were studying a phenomenon that isn't easily defined. What makes a team predictable? Does a team have to win or lose as predicted prior to the game to be predictable? When does one move from being predictable to unpredictable? Many of these questions are not easily answerable. Many questions on your work may not be fully definable. Even so, sometimes they can still be analyzed.

When I got my result, I had to find a means of analyzing its effectiveness. While looking at the results that uncovered the promise of the method, I can also point to other research where our initial intuition of promising results was wrong. In fact, even here, it took a variant of the method to be as strong as we wanted.

As you look at the world, noise will obscure underlying trends. Sometimes, you won't in advance. In other cases, you'll have a sense of what you expect to find. In both instances, you must advance carefully in your findings. Are you finding what you wanted? In sports, did I simply fiddle with the method until it created the results I wanted to see? Is an unintuitive result wrong or is it a breakthrough insight? Often, you simply will not know. You'll need to stop and develop alternative approaches, as we did, that verify the results.

In data analytics, seek and you may find but only because you knew what you wanted to see. In other cases, you will explore and discover, in part, because you did not know what to find.

Keeping these issues in mind can help you to recognize that you have a signal through the noise. It can also help you to carefully analyze an apparent signal that might be present in part of the data but not be a generalized result that can be applied broadly.

So, dig into your data and work hard to find the signal through the noise. If it is hard and even hard to know whether you've actually found something, keep in mind that sometimes, even in that basement, I had to really work to find the broadcast of the baseball game. In the end, it was worth it but it wasn't always immediate.

19.5. Consumable Analytics

Despite being interesting for its sheer magnitude of bits, SportVU data, as discussed in Section 3.2, in raw form, without further processing, was limited. Does it help us to know that Stephen Curry was 15 feet behind the three-point line at 4:24 in the third quarter of a game? Possibly. If you additionally find that he made a three-pointer while covered closely, you have more information. If you compare this type of shot against every game he played this season and find a trend, you have more information. Note how the data makes insight possible but analysis of the data leads to insight.

At the core of much of mathematics is numbers. Data, although the basis of much of sports analytics, is generally not inherently insightful. The analyst must know who will view the analysis, the decision or action that will result from seeing the analysis, and the timeline of the person receiving the work. When factors such as these are integrated into the work, consumable analytics can result. This isn't a guaranteed result of analysis. As an academic, I often hear concerns that mathematics produced from academia can be esoteric, even though interesting. The ability to communicate mathematics helps underscore its importance and relevance to others.

How do we keep from producing narrowly defined work? In sports analytics, first we communicate with whomever will make decisions from our analysis. Keep in mind that a player has different needs from a coach. A team has different interests than those of a fan. Knowing who will receive the work impacts what we analyze and create. Second, we should learn the types of decisions being made from the work. Will a coach use our analysis to help strategize for a game? Will the work be viewed in a few seconds at halftime to devise corrections to game time play? Will a business use our work to directly make a decision or will the work be integrated into other findings before a company decision is made?

Work that's interesting but not actionable generally isn't helpful. Consumable analytics are the goal of my sports analytics group. Just like preparing meals, analytics must be to the liking of the person consuming it. Even more, the meal must fit the setting. A four-course meal isn't appreciated when you have to grab a quick bite between events. A chicken dinner is inappropriate for a vegetarian.

Memorable meals often result from careful planning. From writing the shopping list to creating the meal, there are steps to a dinner. In the same way, you must collect data accurately. You must save it and have it easily accessible. Further, a meal results from good ingredients. Data analysts often talk about rich datasets.

Let's return to our example of SportVu. The NBA invested in collecting spatial tracking data due to its great potential. Even better was the insight the data could quickly offer. If you have the (x, y) position of a player throughout a game, we've seen how we can calculate distances and velocities. We can also splice this data further. Of players who averaged 30 or more minutes per game, Stephen Curry led the league averaging 5.24 MPH on offense in 2017–2018. Note that this analysis came from Second Spectrum as the NBA contract changed from SportVu in 2016.

The NBA saw the impact of insights from SportVU data on their website. Soon after the 2013 season started, the NBA started offering SportVU statistics. The NBA League office stated, "From double-digit web traffic growth for NBA.com/Stats to positive fan email and social media interaction, our fans have shown great interest in the data and expressed how it helps with their analysis and research of the game of basketball."

Do you want consumable sports analytics? Begin with rich datasets. Then, think beyond the numbers. A vital ingredient is insightful questions. The answers, when found, must fit the end user. All these stages are needed, just like shopping, having a good recipe, and cooking the meal itself result in a good meal. Remember that consumable analytics often can't be just whipped up in the analytics kitchen. When appropriate effort is given to an analytics task, you can have actionable results from data relevant to your needs.

19.6. Gaming the System

In Chapter 13, we looked at a cipher that was considered secure, until it wasn't. Note that if a code is cracked, this fact may not be shared since the ability to decipher can be exploited. Exploiting how something works extends beyond ciphers. Let's look at an example related to Google and Internet search.

It's December 2010 and you decide to buy a new tablecloth for a forthcoming holiday party. As many do, you turn to a search engine by typing the word "tablecloth" into Google. Moving to holiday gift buying, you search on dresses in Google. In both cases, the top result is JC Penney, seeming to indicate the ease of one-stop shopping. So, you continue your surfing by searching on "Samsonite carry-on luggage". Again, JC Penney is the top result. Think of the significance of this result — JC Penney outperformed Samsonite.com on a search for Samsonite luggage.

Unknown to you at the time, you could have searched on a large number of items such as "area rugs", "skinny jeans", "home decor", "comforter sets", "furniture", "tablecloths", or even "grommet top curtains". In every case JC Penney would top the list of results returned by Google. Other retailers like Belk, Macy's, Amazon, Nordstrom, or Walmart would appear lower in the list.

Was JC Penney's web pages appearing before Samsonite's in a Google search all that significant? First, think about your own behavior with searches. How often do you click on pages down the list of results returned by a search engine like Google? Many people don't. *The Wall Street Journal* reported that 39% of web users look at only the first search result returned from a search engine and 29% view only a few results. Appearing at the top of Google search results can have a direct impact on sales.

JC Penney's place in Google search results caught the eye of *The New York Times*. The newspaper contacted Doug Pierce of Blue Fountain Media to take a closer look. Pierce's research suggested that Penney's had used knowledge of Google's underlying algorithm to raise the profile of their website on all such queries.

There are literally billions of web pages that are indexed as part of Google's search results. How can a system of this size possibly be gamed? The key is knowing how Google analyzes the web to form its search results.

As we saw in Section 5.4, Google's search engine quantifies the quality of a web page. Simplifying things, if two web pages are equally relevant to a query, then the page with higher quality is listed first. Internet links can be considered recommendations. If your web page has a link from a high-quality web page, then your page is measured as having higher quality. This is a key to higher Google rankings. If you can get links from high-quality web pages, you move up in Google rankings.

Gaming the system to raise your place in the rankings is possible, and legal, but does have risk. If Google finds that you manipulated the system, they can punish you, which ultimately results in a drop in the rankings with a possible major impact on business. Is this fair? That's exactly what created a legal stir in 2002 and 2003. SearchKing.com appeared high in Google's rankings. They would share their high rank (by linking) to their clients' web pages. Suddenly, in August 2002, SearchKing.com's quality measure dropped from 8 to 4 and then from 2 to 0, resulting in the web pages of SearchKing.com's clients essentially disappearing from Google searches.

As a result, SearchKing sued Google for $75,000. The court ruled against SearchKing deciding that Google's results are basically a matter of judgment. Essentially, we choose whether or not to listen to Google's opinion. Consequently, a business can sell links or pay to improve a web page's search engine ranking. While all this can be done, it does risk the wrath of Google.

Gaming Google isn't restricted to business objectives. Step back to November and December of 2003. If you put "miserable failure" into Google, the official White House Biography of the President was returned as the highest ranking query result. Keep in mind that the words "miserable failure" were nowhere on that web page.

This became known as a Google bomb. The architect was George Johnston. Putting together a Google bomb was relatively easy and took about a month. Johnston needed web pages linking to the White House biography of the President with an agreed-upon anchor text, which is the hyperlinked text that you click to go to the linked web page. For the "miserable failure" project, of approximately 800 links that pointed to the President's biography, only 32 linked with the words "miserable failure."

If you searched on "miserable failure" in November to December of 2003, Google returned the White House biography for President George W. Bush as the top result. By January, a "miserable failure" query returned results for Michael Moore, President Bush, Jimmy Carter, and Hillary Clinton in the top four positions. The top result for this query kept switching as the web savvy engaged in this act of Internet politics.

Raising a web page's ranking or setting a Google bomb depends on knowing details of Google's algorithms. Such a knowledge can come from Google's technical and marketing releases and their patents. Releasing such information can lead to effective and important publicity. Patents can play an important role in business. Yet, a knowledge of Google's methods led to the "miserable failure project" and diluted search results in which JC Penney topped Samsonite on its own luggage. This underscores a crux in data analytics. Sharing information on algorithms and methodology allows for transparency. At the same time, the information can allow outside sources to game the system.

Google battles against hackers as gaming their system compromises the integrity of their results. Algorithms to rank items, recommend products, or categorize data are among the many ways we sift through today's deluge of data. Google's rank algorithm distinguished it from other (much more well-known at the time) Internet companies. Giving a sense of a business' algorithms helps establish its place as a pioneer. Yet, making ideas public can have its costs. The balance between transparency and privacy is ever-present.

19.7. Coachable Results

Just because something doesn't work doesn't mean it won't work. Sometimes, a small shift can transform an analytic from interesting to very valuable. Let's see this in sports analytics.

The office is stuffy after hours of discussion and analysis. How can outcomes be improved? Can shortcomings be foreseen and minimized? What preparations should be changed to meet overall goals? Scraps of paper contain scribbles of ideas. A smudged whiteboard has an eclectic outline of the day's conversations.

Is this scene a familiar one? Can you think of a time when you or others in your office enacted such moments? Are you a basketball coach? Why do I ask? The office I'm describing belongs to a coach of the men's basketball team at Davidson College as they analyze team and player performance.

I speak nationally and internationally about my work in data analytics. A popular topic is my work in sports. One point comes up in most talks — produce coachable results.

What is a coachable result? To answer this question, let's discuss how my collaboration with my college's coaches began. In the fall of 2013, a group of four undergraduate math majors and I met with coaches of the

Davidson College men's basketball team called the Wildcats. We decided to explore the potential of our group supplying analytics for the team. For the next two months, we met weekly to hone our work to their needs.

This was an exciting step for our group, which now calls itself Cats Stats. We quickly produced those results that interested us and we presented them. Some of the work led to analytics we supply to the team. Other results were met with a very simple remark, "This isn't coachable." Said another way, we discovered an interesting insight but not something actionable.

Hearing that an analytic isn't coachable triggered a conversation. In order for us to supply helpful results, we had to understand the coaches' needs and goals. They wouldn't mine through the data. They'd use the results as part of their larger work. So, we asked why a result wasn't coachable. Through their answers, we learned more about basketball in the context of their coaching, which made us better data analysts for the team.

In time, we grew to better recognize and produce coachable results. Even now, we do not assume we fully understand the coaches' needs. When we produce any results, we take them to the coaches for their input. Sometimes, we present failed results or less interesting statistics for their feedback.

Let's consider an example after my group produced two new analytics. Given the new work, we set a meeting with the coaches. We thought that one of the analytics was helpful and the other was not. In the meeting, the coaches were excited about both, especially the insight that we thought wouldn't help. Again, this was an opportunity for our group to grow in our understanding. So, I immediately asked, "How is this coachable?" The coaches explained how looking at our work could influence where they might direct the ball in a game or how they might instruct players to move with the ball. We had a coachable result that we almost discarded.

This connects to an important part of our work. How do we know we have a coachable result? Simply put — the coaches tell us. Sometimes, a coach only has a sense that our work will help. In such cases, we supply the analytics and have follow-up conversations to change or refine the work.

In the spring of 2015, *The New York Times*, fivethirtyeight.com, and NPR's *All Things Considered* reported on Cats Stats. As I listened to the students answer reporters' questions, a common theme emerged — produce coachable results. It is the focus of our efforts.

By supplying actionable information, the coaches can use our work in theirs. When we began, we hoped to find analytics that would help the team. Now, we have analytics for the next game on their desks when the current game is done. We are part of the team — from the bench and sometimes the bleachers or dorm room.

Sports analytics can have lessons beyond sports. This isn't entirely surprising as sports encompasses elements of life off the court. As famed announcer Howard Cosell said, "Sports is human life in microcosm."

Are you looking to provide insights via data? To whom are you providing them; what do they consider actionable? If you don't know, a conversation may serve to propel your work forward. If you do know, direct communication with those to whom you'll supply insight, via data analytics or otherwise, can help uncover those hidden gems, which you possibly might see and not recognize as valuable. If you think of yourself as working for a team and focus on finding coachable results, you may find a winning combination of insight.

Between Chapter 10 and this chapter, we've seen a variety of lessons from analytics and mathematics. Keep an eye on your journey as you will likely find your own lessons.

Chapter 20

Getting in the Game

The clock is ticking as the number of pages left to read are getting thin. So, let's conclude with some final tips on getting into the game of sports analytics. I'm a professor of mathematics and computer science at Davidson College in North Carolina. My work specializes in sports analytics. Professional sport teams and national media outlets call to get help from my students and me. Sounds like a dream job? It is to me. So, what tips might I have to lead you into a dream job? It's a question often asked by students who visit my office, and this closing chapter contains my responses to help you uncover your answers.

Tip 1: Know your position and play it well. I enjoy sports more for its athleticism and drama than the statistics. So, I can't tell you the stats of any current baseball or soccer player. I'll miss a game to play with my kids or walk with my wife, although my wife isn't always willing to miss a game. Still, professional sports teams call for help, and I lead a group of students who offer analytics to our college coaches. Did you expect me to know a battery of stats in every sport that I've analyzed? I don't. I won't. I analyze the numbers, which is where I excel. I played sports, so I do have sports knowledge. In fact, our sports analytics group has both students who know very little about sports and others who readily state handy statistics. We design our team this way. A student with considerable domain knowledge (as it is called) can help deepen our work. A student who loves math but not sports might better accept results that fall outside conventional knowledge. So, move in the direction of your strengths, which comes, in part, from knowing your limitations.

Tip 2: Late round draft picks can have exciting careers. In the beginning, I would not have been someone's first choice for sports analytics research. Prior to working in this field, I researched numerical partial differential equations. My shift into sports analytics came via research in ranking. After creating some new ranking methods with my collaborator Dr. Amy Langville, we tested our new work by using the algorithms to create a handful of brackets for March Madness. Our predictions for the basketball tournament outperformed over 97% of over 4 million brackets, which launched me into sports analytics. Now, every March, my phone rings with requests for TV, newspaper, and radio appearances. My work in March Madness propelled me into working with our college teams, which, in turn, led me to work with teams in the NBA, NFL, and NASCAR.

Tip 3: It takes skill and luck to play sports and life. Your skill will move you forward. Nonetheless, randomness and serendipity will also affect your journey. Said another way, you won't always know what an outcome of your work will be. I've given talks where someone in the audience was from a professional team or business who later collaborated with us. I've worked with students from other colleges that later worked in professional sports and helped my group. Whether guided by skill or luck, today's challenge can be a key stepping stone to tomorrow's memorable success.

Tip 4: Have a game plan and take timeouts to re-plan. Who do you want to be? Having a focus on who you want to be can help you make choices as opportunities arise. This is a personal decision and one that can evolve. So, getting personal, two main things drive my professional life. First, I want to work with students and see them become independent creators of knowledge. Second, I want to be active in my family. My professional plate is full, but I carefully protect my family time. Sports analytics achieves these goals. Students actively co-lead the sports analytics group, meeting with coaches and professional teams. Further, shifting into analytics research offered local collaborators, which simplified our family life, especially in the summers. I love my job because the work is fulfilling professionally and at home.

Tip 5: Play to win. Your career impacts your quality of life. You'll spend a lot of time in your job. To me, one guiding principle of any decision is whether I can see it as a win, regardless of how it materializes. I generally don't gamble on an outcome being rewarding. I expect it to be. I simply don't know exactly how that will happen.

What's your dream job? Given the topics of this book, you might expect me to answer sports analytics or, at least, mathematics. In fact, my answer for you is much broader. Maybe your dream job is waiting for you — you haven't seen it yet! What a dream job can be depends on you and what you are today may differ, in large or small ways, from what you are tomorrow. To begin this chapter, I noted that I work in sports analytics. I anticipate that this will be the case for a while, possibly a long while. Yet, I don't know and can enjoy that unknown. Life, in a way, is its own research question. Dive in. Study. Get confused. Discover and keep exploring. I think this allows you to have a career that isn't a game but can definitely be defined as winning.

So, the final buzzer is sounding on this book. Our X Games have concluded. We've covered a number of topics in sports analytics and used those to springboard into questions in mathematics. Your mathematical training can inform any direction you might go with data. As you dive into the numbers, you will bump into areas of mathematics to explore and study. This book is one huge practice field of sports analytics and mathematics. Now, it's your time and your turn to go out, find your areas of interest, and play with sports analytics and mathematics.

Bibliography

[1]   Adams, T. And the Pulitzer goes to...a computer. https://www.theguardian.com/technology/2015/jun/28/computer-writing-journalism-artificial-intelligence, June 2015.

[2]   Antonick, G. Math Madness. https://wordplay.blogs.nytimes.com/2015/03/30/chartier/, March 2015.

[3]   Badenhausen, K. Full list: The world's highest-paid athletes 2018. https://www.forbes.com/sites/kurtbadenhausen/2018/06/13/full-list-the-worlds-highest-paid-athletes-2018/#54531e917d9f, June 2018.

[4]   Barber, S., and Chartier, T. Bending a soccer ball with CFD. https://archive.siam.org/news/news.php?id=1154, August 2007.

[5]   Beckham, J. Analytics reveal 13 new basketball positions. https://www.wired.com/2012/04/analytics-basketball/, April 2012.

[6]   Bergman, J. The world's 10 most tweeted moments. https://www.cnbc.com/2011/10/19/The-Worlds-10-Most-Tweeted-Moments.html?slide=10, October 2011.

[7]   Bosch, R. *Opt Art: From Mathematical Optimization to Visual Design.* Princeton University Press, 2019.

[8]   Bouzarth, E., and Chartier, T. March Mathness in the linear algebra classroom. *IMAGE 52* (Spring 2014), 6–11.

[9]   Brown, L. Lebron James gives funny math lesson after game 1 (video). https://larrybrownsports.com/basketball/lebron-james-math-lesson-video/190859, June 2013.

[10]   Burger, E. B., and Starbird, M. P. *Coincidences, Chaos, and All that Math Jazz: Making Light of Weighty Ideas.* W.W. Norton, 2005.

[11]   Chartier, T. *Math Bytes: Google Bombs, Chocolate-Covered Pi, and Other Cool Bits in Computing.* Princeton University Press, Princeton, 2014.

[12]   Chartier, T. What's your problem. *Journal of Corporate Accounting & Finance* 26(September/October) (2015), 95–96.

[13]   Chartier, T. A bit of humanity in data. *Journal of Corporate Accounting & Finance* 27(January/February) (2016), 97–98.

[14]   Chartier, T. Embracing imperfection. *Journal of Corporate Accounting & Finance* 27(May/June) (2016), 67–68.

[15]   Chartier, T. Miss to make it. *Journal of Corporate Accounting & Finance* 27(6) (2016), 87–88.

[16]   Chartier, T. Step into unknown. *Journal of Corporate Accounting & Finance* 28(November/December) (2016), 78–79.

[17]   Chartier, T. Vertigo over the seven V's of big data. *Journal of Corporate Accounting & Finance* 27(March/April) (2016), 81–82.

[18]   Chartier, T. Graphic stories. *Journal of Corporate Accounting & Finance* 28(3) (2017), 82–83.

[19]   Chartier, T. To hold infinity and beyond. https://www.huffpost.com/entry/math-in finity-_b_1380813, December 2017.

[20]   Chartier, T., and Abbett, M. Cameras, computation, and big data offer new insight in sports. https://sinews.siam.org/Details-Page/cameras-computation-and-big-data-offer-new-insight-in-sports, February 2014.

[21]   Chartier, T., and Ashworth, L. "Be like Mike" on the court — and Bill James in the classroom. https://www.huffpost.com/entry/be-like-mike-on-the-court_b_43 67961, January 2014.

[22]   Chartier, T., Harris, J., and Hutson, K. A probable Super Bowl. https://www.huff post.com/entry/super-bowl-scoring-probability_b_2593151, January 2013.

[23]   Chartier, T., Harris, J., and Hutson, K. To swing or not to swing. *Math Horizons* September (2014), 16–17.

[24]   Chartier, T., and Hickey, T. The red zone exists! Just not where you might think. https://www.huffpost.com/entry/the-red-zone-exists-just-_b_8939020, January 2016.

[25]   Chartier, T., Kreutzer, E., Langville, A., and Pedings, K. Sports ranking with nonuniform weighting. *Journal of Quantitative Analysis in Sports* 7(3) (2011), Article 6.

[26]   Chartier, T., Teal, S., and Stanek, M. Majorly expecting luck in baseball. https://www.huffpost.com/entry/majorly-expecting-luck-in-baseball_b_599 20ccde4b063e2ae05820c, August 2017.

[27]   Colley, W. N. Colley's bias free college football ranking method: The colley matrix explained, 2002. https://www.colleyrankings.com/matrate.pdf.

[28]   Conneller, P. World Cup winner predictions by AI machine learning computer at odds with bookies. https://www.casino.org/news/world-cup-winner-predictions-by-ai-computer-at-odds-with-bookies/, June 2018.

[29]   Costandi, M. How goalkeepers can use an illusion to save penalty kicks. https://www.theguardian.com/science/neurophilosophy/2014/aug/18/muller-lyer-illusion-goalk eeping, August 2014.

[30]   DiLella, C. For the cost of a 30-second Super Bowl ad, you can buy a castle in Kansas City or a Miami condo—take a look. https://www.cnbc.com/2020/01/31/photos-wh at-5point6-million-buys-in-san-francisco-kansas-city-miami.html, January 2020.

[31]   Fox, A. Complete outfield dimensions. https://community.fangraphs.com/complet e-outfield-dimensions/, January 2015.

[32]   Gardner, M. *Mathematics, Magic and Mystery*. Dover Recreational Math. Dover Publications, 1956.

[33]   GroupLens. Movielens. https://grouplens.org/datasets/movielens/, 2020.

[34] Haglund, D. Does basketball actually have 13 positions? https://slate.com/cultur e/2012/05/13-positions-in-basketball-muthu-alagappan-makes-the-argument-wi th-topology.html, May 2012.

[35] Hall, A. On an experimental determination of π. *Messenger of Mathematics* 2 (1873), 113–114.

[36] Henrich, A., MacNaughton, N., Narayan, S., Pechenik, O., Silversmith, R., and Townsend, J. A Midsummer Knot's Dream. *The College Mathematics Journal* 42(2) (2011), 126–134.

[37] Hruby, P. Can Oregon's placard code be broken? http://www.espn.com/espn/page2/ story/_/page/hruby%2F110107_oregon_ducks_signs/sportCat/ncf, January 2011.

[38] Jogalekar, A. Chocolate consumption and Nobel Prizes: A bizarre juxtaposition if there ever was one. https://blogs.scientificamerican.com/the-curious-wavefuncti on/chocolate-consumption-and-nobel-prizes-a-bizarre-juxtaposition-if-there-ev er-was-one/, November 2012.

[39] Kain, E. Madden NFL 19 predicts the winner of Super Bowl LIII. https://www. forbes.com/sites/erikkain/2019/01/28/madden-nfl-19-predicts-the-winner-of-sup er-bowl-liii/#5b9ced4c524f, January 2019.

[40] Kamp, J., and Masters, R. The human Muller-Lyer illusion in goalkeeping. *Perception* 37(02) (2008), 951–954.

[41] Langville, A. N., and Meyer, C. D. *Who's #1:The Science of Rating and Ranking Items*. Princeton University Press, Princeton, 2012.

[42] Layton, J. America's pastime is based on impossible math. https://nypost.com/20 16/10/05/americas-pastime-is-based-on-impossible-math, October 2016.

[43] Loyd, S. *Sam Loyd's Cyclopedia of 5000 Puzzles, Tricks and Conundrums with Answers*. Ishi Press, 2007.

[44] Massey, K. Statistical models applied to the rating of sports teams, 1997. https:// www.masseyratings.com/theory/massey97.pdf.

[45] Moore, D. S. *The Basic Practice of Statistics*, 3rd edn. W.H. Freeman and Co, New York, 2004.

[46] News Service, N. Randomly-generated "scientific paper" accepted. *New Scientist Magazine 2496* (April 2005), 6.

[47] Oaks, O. Ten Twitter facts on the social media giant's 10th birthday. https://www. campaignlive.co.uk/article/ten-twitter-facts-social-media-giants-10th-birthday/13 88131, March 2016.

[48] Paine, N. Why baseball revived a 60-year-old strategy designed to stop Ted Williams. https://fivethirtyeight.com/features/ahead-of-their-time-why-baseball-revived-a- 60-year-old-strategy-designed-to-stop-ted-williams/, October 2016.

[49] Perez, S. Twitter's doubling of character count from 140 to 280 had little impact on length of tweets. https://techcrunch.com/2018/10/30/twitters-doubling-of-ch aracter-count-from-140-to-280-had-little-impact-on-length-of-tweets/, October 2018.

[50] Pillsbury, M. The hot hand and other ways to relate math in sports. https://charlo tteteachers.org/wp-content/uploads/2014/02/Mpillsbury_final-unit_11-26-13.pdf, November 2013.

[51]   Pomeroy, K. The Possession. https://kenpom.com/blog/the-possession/, April 2004.

[52]   Posnanski, J. The Boudreau shift. https://joeposnanski.substack.com/p/the-boudre au-shift, July 2014.

[53]   Prasolov, V. *Intuitive Topology*. Mathematical world. Universities Press (India) Pvt. Limited, 1998.

[54]   Schreiber, S., Smith, K., and Getz, W. *Calculus for The Life Sciences*. Wiley Global Education, 2014.

[55]   Schwarz, A. Professor puts a face on the performance of baseball managers. https://www.nytimes.com/2008/04/01/science/01prof.html, April 2008.

[56]   Singer, Z. Topological Data Analysis — unpacking the buzzword. https:// towardsdatascience.com/topological-data-analysis-unpacking-the-buzzword-2fab3 bb63120, June 2019.

[57]   Solomon, N., and Flindt, R. *Amazing Numbers in Biology*. SpringerLink: Springer e-Books. Springer Berlin Heidelberg, 2006.

[58]   Staff, S. Bizarre play cards in college football. https://www.si.com/college/2014/1 0/02/bizarre-play-cards-college-football, October 2014.

[59]   Stover, C., and Weisstein, E. W. Polar coordinates. *MathWorld — A Wolfram Web Resource* (2020). https://mathworld.wolfram.com/PolarCoordinates.html.

[60]   Stromberg, J. Why do we blink so frequently? https://www.smithsonianmag.com/ science-nature/why-do-we-blink-so-frequently-172334883/, December 2012.

[61]   Thorn, J. One-hit wonders. https://1927-the-diary-of-myles-thomas.espn.com/on e-hit-wonders-49aca2a50ed3, September 2016.

[62]   WashPostPR. The Washington Post leverages automated storytelling to cover high school football. https://www.washingtonpost.com/pr/wp/2017/09/01/the-washing ton-post-leverages-heliograf-to-cover-high-school-football/, September 2017.

[63]   Watt, P. The "Little Planet" effect. http://codeofthedamned.com/index.php/the-litt le-planet-effect, May 2016.

[64]   Weisstein, E. W. Chernoff face. *MathWorld — A Wolfram Web Resource* (2020). https://mathworld.wolfram.com/ChernoffFace.html.

[65]   Wilson, R. Patriots have no need for probability, win coin flip at impossible clip. https://www.cbssports.com/nfl/news/patriots-have-no-need-for-probability-win-co in-flip-at-impossible-clip/, November 2015.

[66]   Witt, J., Linkenauger, S., and Proffitt, D. Get me out of this slump! Visual illusions improve sports performance. *Psychological science* 23(03) (2012), 397–399.

[67]   Witt, J. K. Get me out of this slump! Visual illusions improve sports perfor-mance. https://www.psychologicalscience.org/news/releases/get-me-out-of-this-slump-visual-illusions-improve-sports-performance.html, March 2012.

Index